Town Branch

Writing Collection

Second Edition | 2016–2017

University of Kentucky Department of
Writing, Rhetoric, and Digital Studies

VAN-GRINER

Town Branch Writing Collection

Second Edition
University of Kentucky Department of Writing, Rhetoric, and Digital Studies
Fall 2016–2017

Printed in the United States of America
10 9 8 7 6 5 4 3 2 1
ISBN: 978-1-61740-382-8

Van-Griner Publishing
Cincinnati, Ohio
www.van-griner.com

CEO: Mike Griner
President: Dreis Van Landuyt
Project Manager: Maria Walterbusch
Customer Care Lead: Julie Reichert

Ridolfo 382-8 Su16
170227

Table of Contents

ACKNOWLEDGMENTS

On behalf of the Department of Writing, Rhetoric, and Digital Studies (WRD), we thank the following faculty and graduate students for their contributions to this book: Jeff Rice, Jenny Rice, Brian McNely, Megan Pillow Davis, Joshua Abboud, Craig Crowder, Rachel Elliott, Rachel Hoy, Tom Marksbury, Brandy Scalise, Beth Connors-Manke, Jennifer L. Hudgens, Katherine Rogers-Carpenter, Nathan Shank, Deborah Kirkman, and Ashleigh Hardin. We also thank our editor Maria Walterbusch at Van-Griner Publishing for her work on editing and layout design. We especially thank Dr. Tom Shown for his generous donation to WRD. Without his support and encouragement, this textbook would not be possible.

CHANGES IN THE SECOND EDITION

The second edition includes three new chapters on style, public speaking, and creating documentaries as well as a revised chapter on argumentation, a new index, and content and external hyperlink changes throughout the manuscript. As of the second edition, web links by chapter are available at *http://wrd.as.uky.edu/townbranch.* If you have feedback on the second edition that you would like to share, please e-mail *jimridolfo@uky.edu.*

TO THE STUDENT

Dear Student,

On behalf of all WRD faculty, I am excited to welcome you to the University of Kentucky and to WRD 110, 111, or 112. While the WRD course of study satisfies the first-year general education requirement for composition and communication, the Department of Writing, Rhetoric, and Digital Studies (WRD) also offers a B.A. or B.S. in Writing, Rhetoric, and Digital Studies (WRD), the WRD minor, and a minor in Professional and Technical Communication. If any of these areas of study interest you, contact WRD Director of Undergraduate Studies Dr. Brian McNely. His office is 1315 Patterson Office Tower and his e-mail is *brian.mcnely@uky.edu.* In the pages that follow, you will find descriptions of all three degree programs and additional information is available at *http://wrd.as.uky.edu.* We wish you success in your studies at UK.

Sincerely,

Dr. Jim Ridolfo
Director of Composition
1327 Patterson Office Tower

Minor in WRD.

Nearly every profession values strong skills in writing, visual communication, and persuasion. You can develop these skills with a Minor in Writing, Rhetoric, and Digital Studies, and add real value to your résumé.

3 Easy Steps.

1. Declare the Minor by visiting the A&S Advisors in POT 202
2. Take WRD 300 — Introduction to Writing, Rhetoric, and Digital Studies
3. Take 5 more courses: 3 at the 300 or 400 level, and 2 more at *any* level (200 and above)

Have you already taken a WRD course or two, beyond the 100 level? **They count**.
Questions? Ask! — brian.mcnely@uky.edu | @UKWRD | wrd.as.uky.edu

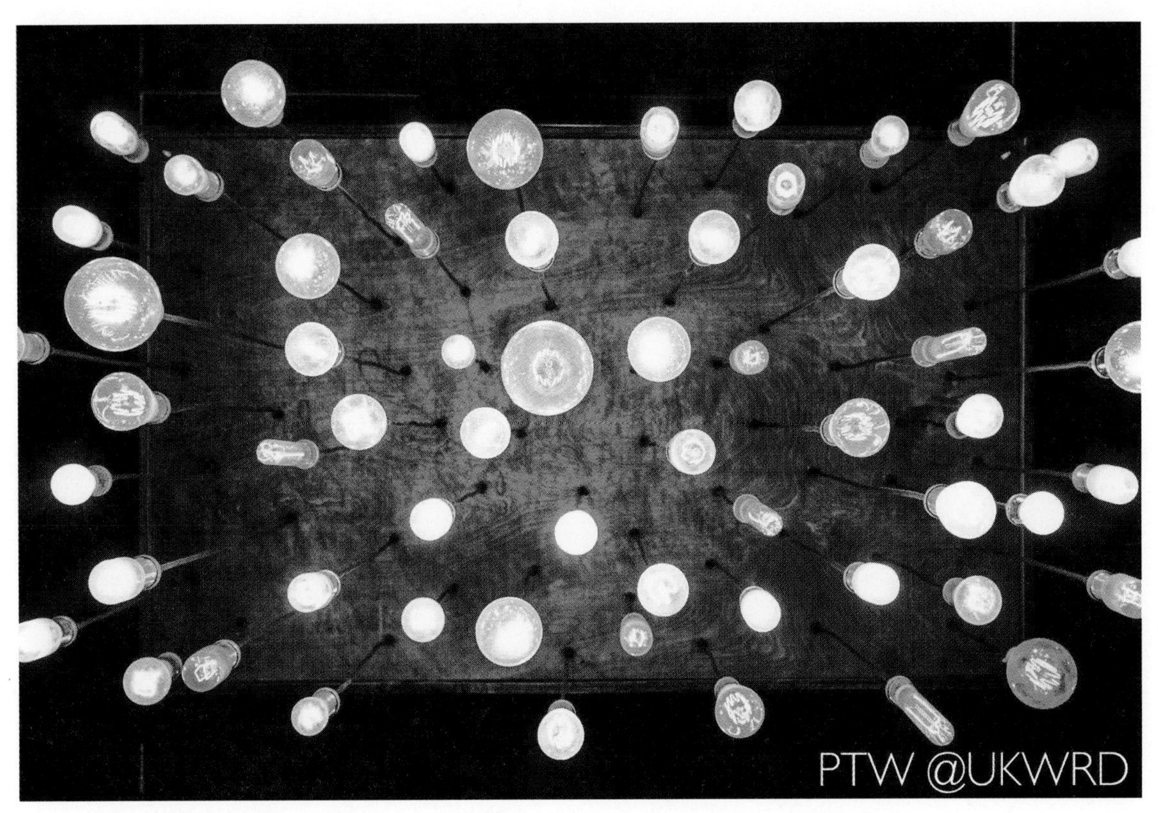

You're filled with bright ideas about how to change the world. Make sure that employers and colleagues hear them, understand them, and *act* on them.

Minor in Professional and Technical Writing.

The most important skills that employers want universities to emphasize are (a) critical thinking and analytical reasoning, (b) the ability to analyze and solve complex problems, and (c) the ability to communicate effectively in writing (Hart Research, 2015). These are *exactly* the skills you'll develop and practice with a Minor in PTW.

3 Easy Steps.

1. Declare the Minor by visiting the A&S Advisors in POT 202
2. Take WRD 204, WRD 300, WRD 306, & WRD 406
3. Choose 2 more courses from WRD 203, WRD 208, WRD 301, WRD 322, WRD 405, or WRD 408

Have you already taken WRD 203 or WRD 204? **They count**. Questions? Ask! — brian.mcnely@uky.edu | @UKWRD | wrd.as.uky.edu

Chapter 1

Rhetoric

A BRIEF HISTORY OF RHETORIC

The word rhetoric has a bad reputation. How many times have you heard someone say, "That's just a bunch of rhetoric." Sometimes people will even use the phrase "empty rhetoric," which is really just another way of saying that rhetoric is all fluff. So, when you tell people that you're studying rhetoric, they might ask you one single question: Why?

If rhetoric were really just empty speech, then it might be a waste of time to study it in college. However, rhetoric is not the opposite of meaningful content. In fact, rhetorical scholars often explain that rhetoric is what gives content its meaning. Rhetoric makes meaning.

To better understand just exactly what rhetoric is and what it does, it might help to begin with a few definitions.

- First-century Roman orator Quintillian wrote, "Rhetoric is the art of speaking well."[1]
- Fourth-century BCE rhetorician Aristotle defined rhetoric as seeing all available means of persuasion.[2]
- Twentieth-century rhetorician Lloyd Bitzer defined rhetoric as "a mode of altering reality, not by the direct application of energy to objects, but by the creation of discourse which changes reality through the mediation of thought and action."[3]
- Twenty-first-century rhetorician Krista Ratcliffe says that "the study of rhetoric is the study of how we use language and how language uses us."[4]
- Finally, a favorite quote for many rhetorical scholars today comes from twentieth-century rhetorician Kenneth Burke: "Wherever there is persuasion, there is rhetoric. And wherever there is meaning, there is persuasion."[5]

What do all of these definitions have in common? Each theorist defines **rhetoric** differently, but they all believe that *language* has some kind of powerful effect in the world. The effects of language are what we might call rhetoric. In Kenneth Burke's description, there is even the suggestion that wherever you find meaning you also find persuasion (and, therefore, rhetoric). That means that every single instance of meaning is a matter of language's powerful effect.

For example, consider these two different descriptions of a snack that I'd like to offer you:

A small cylinder of congealed glucose and gelatin, covered in corn starch.

A light, fluffy treat with a sweet vanilla taste.

Which one would you prefer to eat? Unless you love congealed glucose, you might find yourself more interested in the second version. Both of these descriptions actually refer to the same treat: marshmallows! But the first description definitely would not help to sell many marshmallows. Words like "cylinder" and "congealed" are not usually what we associate with tasty treats. At the same time "light, fluffy" and "sweet vanilla" are quite often words we might connect with deliciousness. Even though the two descriptions are focusing on the same thing—marshmallows—the words we choose can actually create different perceptions for the audience.

In a more serious example, consider how a simple word like "Ms." had powerful effects for many women. Before the contemporary revival of the word "Ms." in the 20th century, women were either identified as "Miss" or "Mrs." The distinction between these two terms marked women as being either unmarried ("Miss") or married ("Mrs."). However, the title "Mr." did not reveal men's marital status. During the 20th century, many women objected to being defined by their husband (or lack of one) in everyday life. To use a neutral title similar to "Mr." When the term "Ms." was suggested, therefore, many women and men welcomed this transformation. With this new title, a person did not need to ask about a woman's marital status before properly addressing her. This tiny word changed the lives of millions of women, and it also helped to create a different perception of women.

This is why rhetorical theorists like to say that rhetoric helps to change reality.

Rhetoric is also sometimes referred to as a form of deliberation. The term deliberation implies careful thought about a complex problem. Juries deliberate during trials, but we also deliberate about a range of issues. Maybe you and your family engaged in deliberation when you tried to decide where to attend college. Did you discuss different school options? Did you find yourself discussing the merits and drawbacks of certain ideas?

In public life, we are regularly involved in deliberations over serious issues. Just take a look at the current news for some indication of what problems, crises, and needs we are now discussing in the public sphere. Some of these deliberations have been going on for years: How should we address environmental problems? How can we end school violence? Should people with terminal illnesses be allowed to have physician-assisted suicide?

We deliberate about these questions because they do not have clear answers. If you and a friend disagreed about the year that Churchill Downs first opened, we would not call this a deliberation. Why not? Because you can do a little research and find that Churchill Downs officially opened in 1875. The disagreement can be settled with factual information that is accessible by everyone. However, not all questions are so easily settled. For those questions that do not have clear answers, we must discuss the merits and drawbacks of different possibilities.

One famous Greek rhetorician named Isocrates (436–338 BC) saw the need for rhetorical deliberation long ago. In his famous text called *Antidosis*, Isocrates writes:

> *For since it is not in the nature of man to attain a science by the possession of which we can know positively what we should do or what we should say, in the next resort I hold that man to be wise who is able by his powers of conjecture to arrive generally at the best course, and I hold that man to be a philosopher who occupies himself with the studies from which he will most quickly gain that kind of insight.*[6]

Isocrates notices that human nature does not contain the power to know everything. Some questions we face do not have definitive answers. But, continues Isocrates, we are not hopeless. Wise people will use their "powers of conjecture" to arrive at the best course. Notice that Isocrates does not say that we will necessarily arrive at the *right answer*, but he does believe we can deliberate with one another to reach the best possible answer.

Recall Aristotle's definition of rhetoric as *seeing all available means of persuasion.*[7] He did not mean that rhetoric is the same thing as persuasion. Rather, he meant that rhetoric is the process of discussing with others in order to see all possibilities in a situation. We use reason, discussion, debate, and persuasion in order to collectively arrive at a response that is fitting. Deliberation, therefore, relies upon powerful usage of language, or what we call *rhetoric*.

WHAT KINDS OF RHETORIC ARE THERE?

A great deal of our contemporary rhetorical theory has very ancient roots. In the fifth century in Athens, the Greek people were trying a new experiment, a form of democracy that required all citizens to participate. As you can imagine, such a widespread participation required citizens to speak in public about legal matters and political matters, regardless of whether they had any type of professional training. For this reason, new schools began to pop up all across Athens. These schools taught rhetoric and oratory, including the art of persuasion. Some of the schools were taught by teachers, known as Sophists. They taught people how to be persuasive when they stood in front of a large group. They later came to have a bad reputation from rival schools, including Plato's philosophical school.

It is Aristotle, however, that we usually turn to for a formal definition of rhetoric. Aristotle was certainly not the first to teach about rhetoric but he was the first to really systematize it and to write it down. Aristotle also believed, much like Isocrates, that rhetoric was necessary because so many of our social and political questions do not have clear answers rather they are questions that we must deliberate with others about. Many of Aristotle's teachings on rhetoric were about how to be persuasive on matters that did not have a right or wrong answer.

Aristotle believed that rhetoricians typically find themselves in one of three situations where rhetorical oratory (or writing) is necessary. These three different situations ask very different kinds of questions, and they require different kinds of discourse and language:

- **Deliberative:** This is the rhetorical occasion that looks toward the *future* in order to ask, *"What should we do about ___________________________?"* Lawmakers very often engage in deliberative rhetoric when they debate about laws that should or shouldn't be passed.
- **Judicial or Forensic:** This is the rhetorical occasion that looks toward to *past* in order to ask, *"What happened?"* Courts of law are great examples of judicial rhetoric, since they are often trying to determine details that have already occurred.
- **Epideictic:** This is the rhetorical occasion that looks to the *praiseworthiness or blameworthiness of someone or something*. Think about giving an award or making a eulogy. Epideictic comes from the Greek word *epideixis:* to shine, to show forth. Award speeches are good examples of epideitic discourse, since they are praising the award receiver.

There are two basic elemental properties of any persuasive discourse: **claim** and **evidence.** Aristotle taught that argument could very broadly be divided into two categories, based on how they use evidence. Following Aristotle, we call these two categories **artistic** and **non-artistic** arguments.

- **Non-artistic argument** is when the evidence supporting the claim is not somehow *invented* by the rhetor. For example, the evidence supporting the claim may be based on eyewitness testimony, the legal language of contracts, and so forth.

> **Example:** *"You owe me $10,000 (claim) because you signed a contract saying you would pay me $10,000 if I built you this doghouse (evidence)."*

- **Artistic argument** is when the rhetor *invents* the evidence used to support the claims. By "invent," we don't mean that the rhetor falsely creates facts and other evidence. Rather, the rhetor is responsible for deciding how to make the argument persuasive to her audience.

> **Example:** *"You should let me borrow your car (claim) because I am very responsible (evidence).*

Most of the arguments we hear and make every day are examples of *artistic arguments*. We may have a claim, but we must consider our audience and decide the best possible evidence that will persuade this particular audience at this particular time.

In the examples above, I do not need to consider what is likely to persuade an audience if I have a contract specifying that you will pay me for services I rendered. On the other hand, if I am about to ask you for a big favor, then I need to consider what is most likely to persuade you. I will choose from a number of possible arguments (the process of rhetorical invention).

This chapter examines types of evidence that is used in *artistic* arguments. We will look at the different qualities of evidence that are used to support claims.

Types of Evidence

Aristotle taught that artistic rhetoric could have three main types of evidence, called **"appeals."** We now commonly call these the appeals of **ethos, logos,** and **pathos.** Any argument uses all three of these appeals, though they each have different effects. Try not to think about these as separate elements that are either present or absent in arguments. They are present in every argument, though they may be emphasized to a greater or lesser degree.

Ethos

Sometimes **ethos** is defined as the character of the speaker or writer. Think about how a writer might seek to support a claim based on her own reputation, background, qualifications, and credentials.

A writer or speaker always creates some kind of ethos. He may draw explicitly upon his personal or professional credentials in order to support his claim. On the other hand, he may fail to establish a very sound or stable reputation. The writer may come across as unprofessional or unreliable. In this case, you might say that his ethos was very unpersuasive.

When you read a writer's text or hear a speaker talk, consider that rhetor's ethos. How does he appear before you? Does he carry weight with you based on how he looks, sounds, or presents himself? Does his language or speech make you feel closely connected to him, or does his speech make him seem distant from you?

Exercise: Thinking about Ethos

Think of how someone's story or background changes how you listen to them. Imagine yourself listening to or reading about international military actions.

Would you weigh the ethos of an author differently if she or he had recently served a tour of duty in the military? Why or why not?

Find an example (TV commercial, magazine advertisement, online advertisement, etc.) of a celebrity or well-known individual selling a product based, at least partially, on his or her ethos. For example, an NFL player selling his shoes is partly using his ethos to help support the claim that these are good shoes. Does the celebrity's ethos convince you to buy the product? Why or why not?

Ethos is something we are always generating, even if we are not aware of it. Another way of thinking about ethos is as a kind of persona. Think about how some entertainers create personas in order to seem more flashy, exciting, bad, or smooth. Even when you don't create entirely new personas for yourself, you are always crafting some kind of public version of yourself for others to consider. Your words, style, sources, quotes, flourishes—these are all part of your ethos.

Exercise: Ethos and Persona

Come up with a new (stage) persona for yourself. Give yourself a new name and style. *How would this persona dress, move, speak?*

Write an introduction that captures this persona in his/her own words and style. Read these introductory pieces aloud (preferably in the verbal style of your new persona)! Discuss whether or not you have put this much thought into your own real persona/ethos in texts that you write.

Ethos is like a set of directions that tell other people how to read you. Your ethos is always sending out these messages, saying quite a lot about who you are, what kinds of things you stand for, how you think, and so forth.

These messages happen even when you don't consciously plan for them! For example, if I wear sweatpants to an interview (just because they're my most comfy clothes, and I want to be relaxed for the interview), I'm saying a lot about myself: *I don't take this job seriously. I don't understand professional behavior. I am probably not attentive to details. I won't be a great employee.* Of course, I'm not trying to send these messages! But, nevertheless, that's my ethos doing the talking.

Ethos and Students: What's one easy way for students to establish a strong ethos with professors?

Pathos

In addition to relying on ethos in making a persuasive argument, rhetors also rely upon the powerful work of emotion and feeling. With each choice of word, tone, or picture, we create a kind of sensation within our audiences. These may be good feelings, sad feelings, confident feelings, proud feelings, fearful feelings, or many other kinds of feelings. Borrowing from the Greek term for *"experience,"* we call this rhetorical power the work of **pathos.**

Although pathos is sometimes translated as emotion, we should probably broaden our understanding of pathos to mean something more like *feeling* and *felt experience*. For example, the experience of boredom might not exactly be an emotion like happiness or sadness, but we have all had the experience of being intensely bored. We have also probably all had an experience of personal shock. Some people describe that experience in terms of embodied feelings: *My heart was racing. It felt like my stomach dropped into my knees. I suddenly felt numb.* These kinds of experiences are not easily encapsulated in familiar terms like "sadness" or "anger." But we can talk about them as felt experiences, or pathos.

The same might be said for the degree and intensity of emotions you feel at different times. A particular argument might make you feel mildly irritated, while the person sitting beside you may feel intense anger. When we talk about pathos, we should therefore consider the *range* of felt experience (including the intensities, emotions, and embodied experiences) that come with those rhetorical choices.

Pathos is a powerful rhetorical tool for persuasion because an audience's emotional state can help move it to certain kinds of action. Think about the old cliché that a speech should always open with a joke. Even though opening with a joke is not always the best advice, the idea is that laughter puts people in a relaxed mood and might ultimately make them feel friendlier toward the speaker. You can also probably think of many examples where authors have used heartbreaking images of sickness, disease, and even death in order to get their audience to donate money or time. A television advertisement for the American Humane Society may use images of abused and lonely animals, all played over melancholy music. The idea is that these images, along with the music and words, may move you to some kind of action (like donating a few dollars to the Humane Society).

Even the most plain, ordinary text is meant to produce *some* kind of responses. Do you remember the feeling that you had when reading your acceptance letter into college? Did the words convey a sense of pride and positivity? If you received any rejection letters, do you remember if they were worded in such a to convey rejection with a sense of regret on the part of the school?

Exercise: Writing to Convey Feeling

In a group, decide how you will create emails that fit the following scenarios:

1. You are a small business owner who is hiring someone to work at the front desk. You had many good applicants, but you were only able to hire one person. You want to craft an email to those people who you could not hire, and you need to tell them that they will not be getting the job. However, you thought these people were very good, and you don't want to discourage them. In fact, you may want to be able to hire them in the future, so think carefully about how to use pathos as a tool for delivering bad news.
2. Imagine that you want to make an argument for giving UK students a full week's vacation leading up to Thanksgiving. You would need to make an argument to the University administration. How might you make your claim and evidence? What kinds of *feelings* would you aim to create for your audience? Would you want to assure them that UK students are *responsible* and *serious*, perhaps as a way of persuading them that you need this time to get important projects done? Would you try to avoid creating feelings of *anger* by insulting past decisions not to cancel classes for the whole week?

Pathos: Always Working

Our words and images always create some kind of experiential feeling, even if we do not intend for those feelings to be felt by our audience.

For example, think about the last time you sat through a lecture that was intensely boring. The speaker probably did not find the material boring, and she most certainly did not want to bore her audience. However, authors do not always have complete control over the way his or her words are experienced by audiences.

Watch a recent speech that your instructor assigns, or one that you find online (TED talks, for example). As you watch the speech, try to determine how pathos is used by the speaker in both her choice of words, her voice, her tone, and even her movements.

Are they effective? Why or why not?

Would they be more or less effective for different audiences or for different occasions?

Feelings versus Logic?

Sometimes people mistakenly claim that pathos should be secondary to other forms of rhetorical appeals. This claim comes partly from certain the post-Enlightenment philosophers who argued that humans are completely rational, and emotions are only a cheap imitation of real thinking. In fact, some philosophers even went so far as to suggest that logic is associated with men, and emotions are associated with women. Because logic was "good" and emotion was "bad," students have historically been encouraged to only rarely (if ever) use pathos in their writing and speaking.

Now, however, we know more about how the brain and nervous system process thoughts and emotions. We have learned that emotions and thought are not mutually exclusive. Neuroscientist Antonio Damasio has written extensively about how thinking and cognition are deeply enmeshed with feeling.

Logos

The third type of rhetorical appeal that Aristotle discussed is one that we now call **logos.** This term sounds like our modern term "logic," but logos is not exactly the same as logic. The Greek word **logos** can also be translated as "word" and "order." You might think about logos as the logic of how claims and their arguments are arranged. Logos is the structure of our arguments.

Rhetorical theorists have argued that we can either structure our arguments using **inductive reasoning** or **deductive reasoning.**

Inductive Reasoning

Inductive reasoning generalizes conclusions from facts or details. You might be familiar with inductive reasoning through your experience with the scientific method. Whenever we practice the scientific method, we make many observations in order to lead up to a conclusion.

As an example of how rhetoricians might use inductive reasoning, consider UK sophomore Emily Markanich's opinion column in the *Kentucky Kernel*, "Today's Dating Is Outdated":

> *Welcome to the age where casual dating has turned into casually hooking up. If you're dating, it's serious, but if you're "friends with benefits" then it's no big*

deal. "Talking" is the equivalent of dating and the mere concept of dating is a frightening, outdated notion in which only our parents willingly participated.

Let's be honest, the times have changed, and not in the most beneficial of ways. We give ourselves away without even considering the consequences of what we are doing to our hearts.

We're either getting our hopes up just to have them dashed by the harsh reality that "he was only into me for my breasts" or "she was only into me for my abs." Or we're hardening our hearts and becoming so callous toward the thought of love that we don't even register the feeling when it presents itself.

This new-age concept of casually "hooking up" isn't casual. It's detrimental. How many of us can say that we know of someone, or have been the person, who's been hurt by casually hooking up?

All of us have, regardless of gender. Sure, it starts out innocent. Just one night to not feel alone, but then that one night turns into a week of nights and only then does the prospect of a date occur.

Some run for the hills, others get their hopes up about a casual fling actually turning into something real for once, only to have those hopes dashed once again. Can't we go back to the social norms to which high school restricted us?

Remember the times when if a guy wanted to "get with" a girl then he would have to ask her out on a date and actually "date" her? Where, "Hey, do you want to come over to night and watch a movie?" wasn't code for, "Hey, want to hook up?"

What I want to know is where the lines became blurred and who is to blame for this. Do we blame the booze, our exes, our parents' marriage, or do we blame ourselves?

And if we only have ourselves to blame, why can't we be the ones to fix it? We teach people how to treat us by how we treat ourselves. I'm not suggesting to turn to abstinence or avoid the bars at all cost. What I am advising is that you show yourself a little respect and treat others with that same respect.

I think you'll be surprised by how far it gets you.[8]

Look at how Markanich constructs her argument. Her main claim—that a culture of more meaningful relationships may also create a more respectful culture—is not explicitly stated at the very beginning. She saves these claims until the very end. "We teach people how to treat us by how we treat ourselves," Markanich writes. The majority of her opinion piece shares small examples of miserable situations that are created by the "hooking up" culture. She describes "dashed hopes," hurt feelings, and misplaced blame. Markanich uses these many examples to build up to her larger claim: we should rekindle older dating practices. By sharing small examples, illustrations, and observations in support of a larger point, Markanich arranges her argument via inductive reasoning.

Deductive Reasoning

Deductive reasoning draws its conclusion from larger principles. Think of deduction as a form of reasoning, or a structure to the argument. One popular rhetorical form of deductive reasoning is called an **enthymeme.** An enthymeme is what we call a **syllogism** that has a missing premise.

Syllogisms are forms of arguments that have two premises and a conclusion.

All men are mortal. (Premise 1)

Socrates is a man. (Premise 2)

Socrates is a mortal. (Conclusion)

In a syllogism, the conclusion is drawn from the previous premises. However, just because a syllogism is constructed properly does not mean that it is true! Consider this logically sound syllogism:

Winter days in Kentucky are perfect for sunbathing.

Today is a winter day in Kentucky.

Today is perfect for sunbathing.

In terms of structure, this syllogism is internally consistent. However, if you know anything about Kentucky winter days, you might doubt the validity of this conclusion! So, be careful of assuming the logically consistent arguments are always true.

You probably don't hear people using full syllogisms in their arguments very often. This is because spelling out all the premises and the conclusion might sound a bit boring or unnecessary. However, imagine what might happen if someone said this to you:

> *There's a blizzard warning for our area tomorrow, so school might be cancelled.*

Did you understand the link between these two statements? If you wrote it out as a syllogism, it might look like this:

> *There's a blizzard warning for our area tomorrow.* (Premise 1)
>
> *Blizzards often cause schools to be cancelled.* (Premise 2)
>
> *School might be cancelled tomorrow.* (Conclusion)

It would have been a little strange to actually state Premise 2 aloud, since you can presumably already figure this part out on your own. Would it have been odd to hear someone say the following?

> *There's a blizzard warning for our area tomorrow, and blizzards often cause schools to be cancelled, so school might be cancelled.*

That unstated premise makes this syllogism an enthymeme. The Greek word enthymeme comes from the words *en*—meaning "to see"—and *thumos*—meaning "the mind." An enthymeme is an argument that literally causes audiences to fill in the missing premise themselves. They "see" it in their minds!

Consider this enthymeme, for example:

> *Congressional candidate Rhonda Weller is not a politician; she's a businesswoman.*
>
> *You should vote for Rhonda Weller.*

What is the missing premise here? The missing premise seems to be that businesswomen (and businessmen) make good congressional candidates, while politicians are not desirable. The author who made this argument did not supply this missing premise, of course. The readers could easily fill in the gaps.

In the example of Rhonda Weller for Congressional candidate, do you think having the reader "fill in the gaps" made this argument stronger or weaker? Is there any kind of strategy to withholding the premise, rather than stating the whole syllogism? Aristotle believed that having audience members fill in the missing premise helped rhetoricians make their case even more persuasive. Why? Because by supplying the missing premise themselves, the audience is actually making the argument for you.

Figure 1. When You Ride Alone, You Ride with Hitler![10]

Exercise: Enthymemes

Arguments are often made through forms other than printed and spoken words. Arguments can also rely upon images to make their case. Enthymemes can also be created through images. Consider the WWII-era advertisement above.

What is the stated premise and conclusion?

What is the unstated premise of this conclusion?

Remember that logos is the structure of various arguments and messages. All texts use some kind of logos, just like every building uses some kind of design. As a rhetorical analyst or author, your job is to pay close attention to the details of that design.

Exercise: Using Ethos, Pathos, and Logos

Imagine that you will be making the argument that all professors at UK must keep assigned textbook costs to $100 or less per course. You will need to make this argument to three different audiences:

- Fellow UK students
- Professors
- University administration

What kinds of ethos would you need to convey to these different audiences? How would you use the powers of pathos in each different situation?

Write a brief outline of your arguments to each of these audiences. Would it be better to use an inductive structure to build your argument? Or, would it be better to use deductive reasoning? What are your stated and unstated premises?

Stasis

Imagine that you and your roommate are having a disagreement about whether you think you should leave the lights and TV going when nobody is home. Maybe you think it is a waste of electricity, but your roommate thinks having the TV and lights on will deter possible robbers. You can imagine yourself standing face to face while arguing about this issue. You are on one side of the issue (it's a waste) and your roommate is on the *other side* of the issue (it's not a waste). Both of you are arguing about the same question—*Is it a waste to leave the TV and lights on when nobody is home?*—but you answer this question differently. You are both in **stasis,** standing face to face on the same issue.

Although it might sound paradoxical, an argument must begin in agreement. That is, people who are participating in a debate must make sure that they are debating the same questions. They will not agree to answer those questions in the same way, of course. People participating in debates and deliberations often have strongly and passionately different answers to the same questions. For example, people who weigh in on the question of how to reduce teenage pregnancy answer in radically different ways:

- Provide better sex education
- Teach abstinence only
- Give teenagers better access to birth control
- Reduce teenage access to birth control
- Teach boys more respect for women
- Have girls play sports when they are young

Even though these answers are very different, even contradictory, they all respond to the same question: *How can we reduce teen pregnancy?*

What happens if you find yourself not in stasis with another person? You have probably had the experience of "arguing past" someone. It can seem like you are arguing two different questions. In fact, chances are good that you probably were arguing two different questions. You were not standing face to face with your interlocutor. Instead, you were each responding to two different questions. Trying to argue with someone when you are not in stasis is a trying activity. For example, some deliberations and arguments about global warming can be quite difficult if person A argues that there is no such thing as global warming, while person B argues that global warming is caused by man-made pollution. In this situation, they are both arguing about global warming, but they are not trying to resolve or respond to the *same question* about global warming.

Stasis theory refers to the place where two disputants "stand"—i.e., the place where they agree to disagree, where both agree something is an issue.

Discovering Stasis Claims

Ancient rhetorical scholars found that most claims could divided into several different types of questions:

Questions of fact: Is it real? Does it exist? Did it happen?

Questions of definition: What is its definition? What is it called?

Questions of quality: Is it good/bad, just/unjust, greater/lesser, etc.? Is it serious?

Questions of cause/effect: What caused It? What effects does it have?

Questions of proposal: What should we do?

These stasis questions were used prominently in ancient Roman law courts, and their form is still used in order to think through legal cases today. In order to proceed with a court case, both the prosecution and the defense must agree to the main question, or claim, being debated. Imagine that I took your umbrella last week without your knowledge, and I suddenly found myself appearing before a judge. Before I am charged with anything, the court may want to first run through the stasis claims in order to decide what argument to make:

Question of Fact: Did I Actually Take Your Umbrella?

If I did not actually take it, then we would stop here. I would argue that I was framed or wrongfully accused, and you would argue that I did indeed take the umbrella.

However, let's say that I admit to taking your umbrella. Then, we would move on to the next stasis question.

Question of Definition: Was the Act One of Theft, or Was It Something Else?

I may admit to taking the umbrella, but I do not want to define it as theft. I want to say that it was simply "borrowing." I was going to return it in perfect shape. However, you might argue that taking without permission is always considered theft.

But, let's say that I admitted to actually stealing the umbrella. It was theft. We can agree on that question, so we'll move to the next stasis question.

Questions of Quality: Was My Act a Terrible Act?

Let's say that I agree to have stolen your umbrella in a moment of sheer thoughtlessness. I was feeling mean or rotten on that day. But I might also argue that taking an umbrella is not a terribly awful act. It did not impact your life in serious ways (beyond having to buy another umbrella), and you were not physically or emotionally harmed. You might not share this opinion, however, and you want to see me spend the rest of my life in prison for theft. In this instance, we would be arguing over questions of quality: How bad was this act?

On the other hand, maybe you agree that the act was mean, but was not terribly horrible. We might want to still ask one more question before moving on to the bigger question of what to do.

Question of Cause/Effect: What Caused This Act? What Effects Did It Have?

You might argue that the monetary price of your umbrella was very insignificant, but the umbrella had been a present from your favorite grandmother. The sentimental attachment to your umbrella is priceless, and you have found yourself depressed for weeks as a result of the theft. Furthermore, even after you got it back, you now find yourself unable to trust people around your umbrellas ever again. In this case, you would be arguing about effects. If I try to argue that such effects are highly improbable from a simple case of umbrella theft, we would be arguing the stasis of cause/effect.

Finally, if we could resolve all of these other questions, we might still find ourselves arguing the question of proposal.

Question of Proposal: What Should We Do?

You may wish to see me pay a fine or even serve some community service as a result of my theft. However, I may argue that an apology (and the returned umbrella) should be sufficient. If we had disagreements about the appropriateness of various proposals, we would be arguing the stasis of proposal.

How Can You Use Stasis Questions to Understand Others' Arguments?

The stasis questions are incredibly helpful for understanding the arguments and claims that others are making. When you read or hear an argument being made, you might ask yourself the following questions:

- Is the author making a **claim of fact:** Is s/he arguing whether something exists or whether it happened?
- Is the author making a **claim of definition:** Is s/he arguing what something should be called or how it should be classified?
- Is the author making a **claim of quality:** Is s/he arguing whether something is good or bad, just or unjust, greater or lesser, etc.? Is s/he evaluating something?
- Is the author making a **claim of cause/effect:** Is s/he arguing about the cause the led to something? Is s/he arguing about the effects that something will have?
- Is the author making a **claim of proposal:** Is s/he arguing about the best course of action in a certain situation?

Once you understand what kinds of claims and questions are being deliberated, you can formulate a more appropriate response to your interlocutor. You can also use this list of stasis questions to reflect and better understand your own claims. Are you in stasis with your interlocutors? Are you all arguing about the same questions? If not, how can you first agree to what your argument will be about?

Exercise: Shifting Stasis

Find a recent opinion piece in your campus newspaper. Use the above questions to try and determine what kinds of stasis claims the author is making. Now try to imagine how the author's argument might change if the main stasis were shifted to another kind of stasis claim.

For example, if the author's main claim is one of definition, how might it change if he or she were to make an argument of quality?

Using Stasis for Inventing Arguments

Not only are the stasis questions useful for analyzing other people's arguments, but you can also use the stasis claims in order to help formulate your own arguments. This is why we say that the stasis questions are a form of invention or **heuristics.** If you know that you will need to make an argument about a certain issue, but you are not quite sure what you want to say about it, use the stasis questions as a way to generate ideas.

The following is an inventional writing exercise that might help to get your thoughts flowing:

Exercise: Invention

1. Sit down with a piece of paper and a timer. Write down your main issue or topic focus in a few words (for example, "veterans returning to school").
2. Spend five minutes generating as many questions as possible under each of the stasis headings. (Use the list below as a guide.) You should come up with questions about your topic—not answers—relating to that particular stasis area. For example, under the "fact" stasis, you will try to come with as many questions about your topic that relate to its "facts" or existence.
3. When you are finished, look back over your questions and circle the ones that seem the most interesting to you. If you would like, you can do another timed writing exercise for each of the circled questions. Go through the whole list again.

Exercise: Invention (continued)

Questions based on stasis areas:

Existence or Facts

- Did ____________ happen?
- How did it happen?
- What are the facts of ____________?
- How did ____________ begin?

Definition

- To what kind of larger class of things or events does ____________ belong?
- What parts of this story need more definition, or does this person's words suggest a new kind of definition?
- How is ____________ different from other things?
- Is the meaning of ____________ misunderstood?

Quality

- How serious of a problem is ____________?
- Is ____________ a good thing? A dangerous/helpful/beautiful/etc. thing?
- How do different people/stakeholders think about ____________? The stakeholders involved may include: ____________.

Cause/Effect

- What are the causes of ____________?
- What are the consequences of ____________?
- What kinds of effects does ____________ have?

Proposal

- What should be done with ____________?
- Should we reconsider how ____________ works?
- What kind of future should ____________ have?
- Does ____________ need any kind of change?

VIDEO RESOURCES

Mayor Jim Gray on the Rhetorical Force of Words

Mayor Jim Gray was elected Mayor of Lexington in 2001. Mayor Gray is also the former CEO of Gray Construction, which is a successful international construction firm. As part of his daily work in the Mayor's office, Mayor Gray uses writing in a number of ways. In this interview clip, he talks about how he chooses particular words for their rhetorical effect.

(https://vimeo.com/103941623)

Mayor Jim Gray on Using Pathos and Emotional Persuasion to Change Perceptions of Place

In this clip, Mayor Gray describes how he uses pathos (emotional appeals) in order to affect how people see a place like Lexington.

(https://vimeo.com/103941621)

Kevin Patterson on the Everyday Importance of Rhetoric

Kevin Patterson manages the Beer Trappe, a craft beer specialty shop in Lexington. Kevin is a native to eastern Kentucky, and he is an active participant in social media discussions of craft beer. As part of his job, Kevin uses rhetorical theory and writing in a number of ways. In this interview clip, he describes the role rhetoric plays in his everyday work life.

(https://vimeo.com/103940588)

Kevin Patterson on the Importance of Storytelling and Perception

In this clip, Kevin Patterson talks about the importance of storytelling and narrative on how people perceive different products.

(https://vimeo.com/103939686)

NOTES

[1] Quintilian, *The Institutio Oratoria of Quintilian*, trans. H. E. Butler (Cambridge: Harvard University Press, 1996).

[2] Aristotle, *Rhetoric*, trans. W. Rhys Roberts (Hazelton: University of Pennsylvania, 2013), I:4:1359.

[3] Lloyd Bitzer, "The Rhetorical Situation," *Philosophy and Rhetoric 1* (1968): 4.

[4] Krista Ratcliffe, "The Current State of Composition Scholar/Teachers: Is Rhetoric Gone or Just Hiding Out?" *Enculturation*, *5,* no. 1 (2003).

[5] Kenneth Burke, *Rhetoric of Motives* (Berkeley: University of California Press, 1969), 172.

[6] Isocrates, *The Antidosis*, (Cambridge: Harvard University Press, 1929).

[7] Aristotle, Rhetoric, I:4:1359.

[8] Emily Markanich, "Today's Dating Is Outdated." *Kentucky Kernel* (Lexington, KY), January 29, 2014.

[9] Ibid.

[10] Weimar Pursell, *When You Ride Alone, You Ride with Hitler!*, 1943, http://www.archives. gov/exhibits/powers_of_persuasion/use_it_up/images_html/ride_with_hitler.html.

Chapter 2

Rhetorical Situation

When we get ready to leave our house and go out into the public, we often do a certain amount of preparation and take into account the nature of the task we are setting out to do. If going for a job interview, we may try to anticipate what questions we will be asked, and we do our best to sound confident and well informed during the interview itself. If we know we are going to a doctor's appointment, we make sure we have a method of payment and insurance info, mentally run down what we need to ask or bring a written a list, and try to arrive to the office right on time. When we do these things, we prepare ourselves to encounter a particular situation—one that hasn't happened yet, but we still know how to plan for—and then navigate that situation to the best of our ability. In doing so, we increase the likelihood that our "mission," whether it be landing a job or picking up a carton of milk, will succeed.

When we communicate, we also have to prepare for the situation in order to increase the chances that our message will be well received by others. Like the diverse errands and activities we complete during our daily lives, acts of communication also have their own specific requirements and contexts. These external factors are what make up a **rhetorical situation,** the phrase rhetoricians use to describe the circumstances in which a rhetorician finds him- or herself whenever there is some need to speak out about some issue or problem.

Lloyd Bitzer, a twentieth-century rhetorical scholar first introduced the concept of rhetorical situations in his well-known essay "The Rhetorical Situation." In that essay, Bitzer offered the following definition:

> *Rhetorical situation may be defined as a complex of persons, events, objects, and relations presenting an actual or potential exigence which can be completely or partially removed if discourse, introduced into the situation, can so constrain human decision or action as to bring about the significant modification of the exigence.*[1]

That is quite a technical definition, but here is one way we might translate Bitzer's words into more familiar language:

> Rhetorical situation may be defined as a combination of certain people, events, objects, and relationships that seem to be related to some problem that can potentially be solved by someone persuading the right people to change the problem.

In other words, every rhetorical situation has an **exigence** (a problem), a **rhetor** (the person speaking out), an **audience** (the people who could change the problem, or the people hearing the rhetor), and **constraints** (things that might limit the rhetor or the audience).

For example, imagine that your professor decides to change the syllabus in the middle of the semester. Instead of taking only a midterm and final exam, you will now have to take two exams and write a ten-page paper. You and your classmates believe this sudden change is unfair, so you might decide to do something about it. Together with your classmates, you politely ask for a meeting with your professor where you can outline the reasons why this change is unfair to all the students in the class.

In this example, the exigence or problem is that you have a very unexpected (and unwelcome) change to your workload. You and your classmates are the rhetors. You will reach out to your audience through your discourse. Of course, you have certain constraints on you: not only is your professor the one who assigns your grades, but he may have been teaching this course for a long time. He may or may not be interested in hearing your points. He may have simply made a mistake when handing out the original syllabus, and he is now very apologetic. Whatever the case may be, it will help to fully understand the rhetorical situation before you and your classmates decide how to approach your professor.

In this unit, we will discuss the elements of the rhetorical situation and how you might use this model as a way to both analyze and produce public arguments. While each rhetorical situation will have its own specific context, there are some common features or elements that will be present in every rhetorical situation. Just as the various shirts in your closet contain different patterns, styles, and textiles—yet all have neck and arm holes—each rhetorical situation will possess the following common features, but remain unique due to the particular context surrounding it. These main elements common to all rhetorical situations are exigence, audience, and constraints.

ELEMENTS OF A RHETORICAL SITUATION

Exigence

A rhetorical situation comes into existence the minute we perceive the need for a problem to be solved. **Exigence** is the term used to describe the impetus that provides such a need. Exigence may take the form of a problem, obstacle, or some other defect or deficiency that the rhetor identifies, wishes to overcome or alter, and that can be ameliorated through discourse.

This last criterion is particularly important because in order for a rhetor's message to be successful, it must be delivered to an audience that could potentially be persuaded by it, on one hand, and it must address a problem or conundrum that is potentially fixable through rhetorical appeal, on the other.

Thus while one could argue rain is a nuisance best confined only to agricultural areas that need it for crop health, such a line of reasoning doesn't qualify as exigence because it would receive little support; endorsing this belief will not actually prevent rain from falling in cities.

Instead, an example of exigence might arise when a spectator attends a tennis match where a participant is injured playing on a wet court. The spectator might recognize that the situation could have been prevented and begin lobbying for a change in rain-delay protocol. Since rain-delay procedures could, feasibly, be altered or amended due to such advocacy, we would consider this motivation for speaking out an example of exigence.

In addition to furnishing the motivation to deliver a message, a rhetorical situation's exigence can have a profound effect on a rhetor's discourse by helping, in part, to determine the appropriate audience to persuade. If we stick with the rain example mentioned above, we could see how the nature of the problem (loose rain regulations can lead to athletes being injured) could influence whomever a rhetor chooses for his or her target audience (members of tennis associations, umpires, tennis fans, professional players, etc.).

No Exigence? No Argument.

If there is no exigence, no perceived problem, then a rhetorical situation seems not to exist. Consider how odd it might be to try and persuade an audience that they should breathe with their lungs.

Of course, there is no real need to make such a persuasive case on this subject, since we breathe with our lungs whether we like it or not. Most of us do not have any options. You could indeed make such a speech, but it's more likely to be an empty exercise than a real rhetorical situation.

Furthermore, when rhetors think critically about their situation's exigence and why it is important, they often discover deeper, underlying issues or problems that add to the urgency and ethos of their proposed course of action. The argument is no longer about a system's failure to protect a particular athlete, on a particular day, playing on a particular court. Instead, the event becomes symbolic of a cultural proclivity to put fans' wishes ahead of competitors' safety or the tendency to be reactive rather than proactive about athletes' health during competitions.

By including the implications behind a rhetorical situation's exigence, you can actually create a stronger message because the various people who encounter that message will often relate more strongly to issues of public concern (which they feel they have a stake in), rather than one isolated incident (which they may feel does not pertain to them). In this way, exigence is important not only because it provides the catalyst for a rhetorical situation to develop, but because it also shapes and contributes to the exchanges between rhetor and audience that comprise rhetorical discourse as a whole.

How Can Thinking about Exigence Improve Your Ability to Persuade Others?

When you want to persuade an audience, your first job is to convince them that there is a problem that needs solving. Your audience needs to believe that there is an exigence that they can do something about. If you can accomplish the following goals before moving into your main argument, you will gain your audience's attention from the very beginning.

From the Very Beginning, Explain What Your Message Is About

As a rhetor, clearly identifying your topic and goal can lead to a more focused message that comes across as polished, refined, and cohesive. It can also help ensure you stay on track during the research and composing processes. In other words, you are trying to provide an anchor for your audience. From the beginning, your audience should have a sense of the general topic you will be addressing.

Explain Why Your Message Is Needed

After anchoring your audience by introducing the general topic, you will need to establish the exigence. By telling them why your message is needed, you are trying to persuade your audience that they have something to lose by not listening to you. They also have something to gain by continuing to give you their undivided attention. It is not enough for you, the rhetor, to feel that there is a problem to be solved. Your audience must also share that feeling. You are responsible for helping to create that shared feeling with your audience.

For example, say we are writing a piece arguing against the proposed natural gas pipeline that, if approved, would span our state and run through eighteen Kentucky counties on its way to the Gulf of Mexico. We may say that our message needs to be heard because the pipeline has not yet been approved and so there is still time to prevent it from being built, or that we need to wait and assess the environmental impact the pipeline's construction could have on our state's historical sites (some of which lie in the pipeline's proposed route).

Such answers are important and should be included in the message. However, we could broaden the reasons *why* this course of action should be pursued by mentioning concerns and issues relevant to people beyond our state's borders—such as land rights and usage, the potential problems of imminent domain and property rights, and the impact of such projects on the environment we all share. Doing so provides even more ways for audience members to connect with a topic and feel strongly about it—important steps toward persuading people to adopt a specific course of action toward or proposed solution to a problem.

Help Your Audience to Visualize What Should Be Accomplished in Response to the Exigence

As you help your audience to understand the exigence they face, you should also help them to understand the goals that can address the problems. Think about what you want to accomplish. Are you proposing a solution that is truly cause for celebration, or will the solution necessitate a certain amount of sacrifice for the greater good or require short-term hardship for long-term benefits? By specifically framing your proposed course of action, your audience will know that you've thoughtfully considered the situation's exigence and the merits and drawbacks of potential solutions. It will help you come across as realistic and well versed in the topic chosen, thus enhancing your ethos.

Audience

Discourse is a two-way street. A speaker or writer may feel moved to create written or spoken messages, but unless these messages are later interpreted by someone (even if that person is the rhetor, like in a private journal), the messages dead end and communication breaks down before it has begun. If communication is stifled and information does not pass from rhetor to listener/reader/viewer, a rhetorical situation does not develop because no discourse is able to occur. For this reason **audience** is a key element of rhetorical situations.

Types of Audiences

There may be several different types of audiences involved in one rhetorical situation. Think, for example, of the wide variety of people who may read an editorial article in a newspaper. There will be those who read the article because they are interested in the topic and also hold a strong opinion about it; these readers may either agree with the author's point of view or be actively opposed to it.

Additional readers will be undecided on the issue and in search of more information before they draw their own conclusions. Other readers will simply "find" the article in the paper and decide to read it on a whim. These readers will live in different areas, be of different ages, and possess different needs (desire for information, entertainment, etc.). For these reasons, it is important to think about who could potentially encounter your message and which methods might serve best when appealing to those you, as the rhetor, hope to persuade.

General Public Audience

As the name suggests, this term refers to an audience comprised of various individuals from all walks of life. It could also be thought of as a generic, "every man/woman" sort of audience base.

Yet what might seem to be a rather straightforward and simple concept on the surface can become more complex when considering the people a particular rhetor might imagine comprise the "general population." No one composes in a vacuum. Instead, all rhetors are influenced, whether consciously or not, by the societal norms characterizing their own specific communities and culture. As a result, some rhetors may assume that the general population is made up of people whose world views, beliefs, and personal background are similar to their own.

While this may sometimes be the case, often the diversity found in an audience of the general public is much greater due to the non-specific nature of the audience pool (one need only read the comments section of any online article to see an example of this). While this large coverage can result in a message receiving more "air play," many who encounter a message made for a general audience may fail to heed its importance because they do not feel the message applies to them, find it off-putting because the message assumes too much, or find that the message simply does not "stand out" from the myriad other messages vying for the general population's attention. For these reasons, many rhetors find that narrowing their audience base can enhance the actual productive power of their messages.

Target Audience

Instead of beginning from a general public *audience*, rhetors are more successful when they develop a more narrowly focused target audience. The term target audience is used to refer to individuals the rhetor *intends* to persuade and desires to speak with. They are the rhetor's target.

Generally, There Is No Such Thing

Look around the room right now (or next time you're in a crowded room). You might well have some things in common with these people. You might all be college students, or you might all be under forty years old. Now think of how much you probably don't share in common with everyone in this room. Some of you might be extremely liberal, while others are extremely conservative. Some of you might have experience with terminal illnesses, while others of you have never experienced a loved one's death. Some of you might love watching reality TV, while others of you absolutely hate TV altogether. Some of you might be very religious and others not religious at all. The point is that even an audience who seems to have a lot in common can actually be extremely different. This is what makes the "general public audience" a pretty unhelpful concept. There is no general public audience. Publics (even just small crowds, like the one you're in) are filled with people of different backgrounds, opinions, and beliefs.

Often, rhetors will formulate their idea of target audience during the early composing and researching stages because anticipating the needs, preferences, and expectations of a target audience can help a message's appeal. For example, if a blog writer wanted to target women aged 24–35 with disposable income, the writer might include cultural references thought to appeal more to women, use "feminine" colors when setting up the site, and utilize a vocabulary familiar to college-educated individuals. Though certainly not an exhaustive list, rhetors will often formulate their target audience with the following criteria in mind:

- What level of familiarity does my target audience have with this subject?
- What age is my target audience (approximately)?
- What type of vocabulary would my target audience be most familiar with (keeping in mind field-specific jargon, "buzz" words, etc.)?
- Are there any cultural preconceptions my target audience might have that I need to be aware of?
- Is this target audience hostile or open to my message?
- What tone and style is my target audience is used to?
- How might my target audience respond to my examples/research/reasoning? What could they object to or be persuaded by?

The Power of an Audience

Initially, we may think that an audience has limited power in a rhetorical situation since the rhetor composes and delivers his or her message to a "passive" group of listeners. Audience members, however, actually can exert quite a bit of influence over a rhetorical situation. Because an audience is made up of dynamic human beings who will use their own intellects, opinions, and communications to respond to a message or its features, an audience has the potential to radically alter a rhetorical situation.

Anyone who has ever witnessed a political speaker shift from opinionated comments to rapidly apologetic back-pedaling has seen a prime example of how an audience's reaction to a message can determine the future course of that message. Because an audience is able to respond to messages with emotion and intellect, they often force a successful rhetor to adapt his or her message by clarifying, elaborating, or otherwise revising the information presented—sometimes even to the point where the initial context of a rhetorical situation is changed (such as when an informational meeting morphs into a debate, or a casual conversation becomes a profession of love).

Exercise: Exigence and Audience

Imagine that you are part of a group that wants the FDA to make experimental drug treatments easier to obtain for patients with terminal illnesses. Together with your group, decide what different types of audiences you might reach out to, and then decide how those different audiences might best be reached.

Discuss how your different target audiences might perceive the exigence (making drug treatments easier to obtain for terminal patients). Will they share your sense that this is an urgent problem that should be addressed? If some audiences might not automatically share your sense of urgency, how can you help to persuade them that this exigence is important?

The relationship between a rhetor and the audience is give-and-take in nature. For this reason, it is wise for any rhetor to try to anticipate the expectations, needs, and possible reactions an audience might have when composing and refining the rhetorical moves employed in his/her message.

Constraints

Anticipating how an audience could interpret your message is one important way to acknowledge and work through **constraints.** Constraints are any external factors, aside from the rhetor and audience, which could impact a message's success. Often, constraints refer to the parameters the rhetor must compose in and the context within which the audience will receive and interpret a message. For example, rhetors often operate under constraints that arise due to genre (a press release cannot be as long as a novel), occasion/social convention (it's best not to curse during a eulogy), or culture (it's important to write in a language the majority of readers know). Likewise, the social and historical context surrounding an audience and rhetor during a rhetorical situation can place additional constraints on the table. We can see the power of such constraints in current debates surrounding the inclusion of rape jokes in stand-up comedy or the resonance of Congressman Bobby Rush wearing a hoodie on the House floor shortly after the death of Trayvon Martin in 2012.

What is going on in the world "outside" of a rhetorical situation matters because, in reality, it is impossible to not bring the "outside world" into rhetorical situations through the various associations, symbols, and preferences we all carry with us every day. It is this inclusiveness that gives rhetorical discourse its potentially persuasive power and creativity, but it also means that we as rhetors and audience members should be aware of constraints and the possible impact they could have on our messages and our interpretation of the messages of others. Doing so not only means being a responsible rhetor or audience member, but also means being a more active participant in the exchanges of information that formulate our lives.

Types of Constraints

Genre Constraints

Genre constraints arise as a result of the particular medium a rhetor is working in. The different types of communication (written, spoken, and visual) have various qualities not shared with other types, and the multiple forms available within these larger categories will also have their specific traits (think rock music versus opera music). If a rhetor wishes to create a message that accords a genre, he/she must be familiar with the distinctive elements that comprise that genre. Genres may also be blurred or combined in order to create new forms of communication, but it is still important to know the constraints that help place a message into a specific genre so that such blurring/blending can be successfully recognized and appreciated for what it is. Genre-based constraints often center upon one of the following:

- Tone
- Visual presence
- Style
- Duration
- Text presence
- Design
- Delivery method
- Purpose/use
- Artistic precedent

Exercise: Genre Constraints

Together with a partner, come up with short speeches that fit both of the following occasions. You might choose one person to be the speechmaker, or you might want to switch off.

- Write a "wedding toast" for your partner. This should be done in the style of a wedding reception toast, in order to praise and celebrate him or her. (Have fun with this speech! Feel free to make up details that might make this speech appropriate for the occasion.)
- Write a short eulogy for your partner (who happily lived to a ripe age of 100 years). This should be done in the style most appropriate to a memorial service or funeral.

How did the different *genres*—wedding toasts and eulogies—constrain what you were able to write? Even though both speeches were praising or honoring your partner, did you notice that the occasions changed such details as tone, delivery, style, and even word choices?

Material Constraints

Material constraints are those constraints that occur due to the materials or equipment needed to compose and deliver a message successfully. They can range from making sure you have enough pens and paper, computer access, or a working microphone to ensuring that you have clear images, are speaking in a venue that can accommodate your visual aids, or can fit your text into the newspaper column space allotted.

Cultural Constraints

Cultural constraints refer to those factors that can influence a message's efficacy and are socially constructed. Cultural constraints may take the form of proprietary "rules," those unspoken guidelines that determine what proper etiquette is during a particular rhetorical situation. They may influence a rhetor's word choice, balance of pathos/logos/ethos, or even a rhetor's choice of dress.

Purposely Breaking Convention—When Constraints Are Undermined

Occasionally, rhetors will intentionally thwart constraints in order to produce a reaction from the audience or to break the boundaries of what a traditional genre is thought to encompass or do.

For example, in July 2013 MSNBC's news anchor Melissa Harris-Perry showed her disapproval of the Texas Department of Public Safety's decision to ban women from bringing feminine hygiene products into the state's capital building during debates surrounding the state's abortion law by wearing a pair of tampons as earrings on the air. Obviously, such "jewelry" does not fit our society's traditional expectations of how public speakers and news anchors should dress when addressing their audience and the action did, most certainly, make some uncomfortable. For Harris-Perry, however, such discomfort was the point. Summing up why she chose to act this way, she said she felt it would help the audience "remember the Texas state legislature said [women] could not bring tampons in when they were going, these women, to in fact stand up for their own reproductive rights."[2] While Harris-Perry's attire drew both praise and criticism, it also kept the political context surrounding her action is the forefront of the public's discussion. Purposely undermining some constraints can, then, be a powerful tool. However, the rhetor who chooses such a route also assumes great risk; the incident could backfire (for example, Harris-Perry was accused by some of merely wanting attention) or alienate members of the rhetorical audience who might have been sympathetic to a more even-keeled approach.

Other times rhetors may choose to eschew some of the conventions of a particular genre or to mix and match across several genres. Many multimedia presentations do this by mixing element of written text, visuals of various types, and other sound components. They may sample various pieces from "well-known" works of art or music, or mesh together genres to create a hybrid (such as a graphic novel). Often this synthesis of styles and genres produces innovative and interesting messages that appeal to audiences. However, adopting such a path also carries risks. If a synthesis is too extreme, an audience may find it jarring or incomprehensible.

Ultimately, it is the job of the rhetor to deduce the constraints most important to him or her, figure out how to accommodate them effectively, and decide whether the risk of breaking some of the constraints outweighs the possible consequences that could arise from such a choice.

RHETORICAL SITUATIONS AND THE WRD CLASSROOM

As you complete the WRD curriculum here at UK, you will have the opportunity to compose in a variety of mediums and genres. Since WRD assignments often include aspects of written, oral, and visual composition, you will become well versed in many of the conventions characterizing these mediums. A keen sense of the rhetorical situation surrounding these assignments, from the parameters the instructor requests for each assignment to the nuances of your own target audience, will help the efficacy and overall impact of your message.

In addition, some assignments you receive in WRD may require you to conduct a rhetorical analysis of someone else's work. Thinking about the context and factors that comprised the rhetorical situation surrounding this person's work may help to facilitate this task. It will require you to use your powers of critical thinking, but if you examine the ways in which a rhetor's work upholds or plays with genre conventions, social expectations, and cultural norms—you may find some interesting information to include in your analysis.

NOTES

1 Lloyd Bitzer, "The Rhetorical Situation," *Philosophy and Rhetoric 1* (1968): 6.

2 Robby Barthelmess, "MSNBC Anchor Protests Texas Abortion Bill by Sporting Tampon Earrings," *News. Mic,* last modified July 23, 2013, http://mic.com/articles/55985/msnbc-anchor-protests-texas-abortion-bill-by-sporting-tampon-earrings/742487.

Chapter 3
Audience

When you write, you always write for someone else. Even if that someone else is your teacher, you still are writing for *someone*.

Writers sometimes forget about audience. In school, writers produce quickly written work and forget about the people who will eventually read that work. As a result of this oversight, our work can sometimes get weighed down with clichés, grammar errors, a lack of organization, a lack of research, and other problems. When these problems appear in our writing, it is usually because we have forgotten to consider our audience. And, unfortunately, if too many of these problems exist, we can assume that the reader will most likely stop reading at some point.

Unless you are writing something private, like a journal or a diary, you can be sure that at some point someone will be reading your writing. Maybe it will be your instructor and your class peers. But, for the rest of your life, your writing will probably also be read by many other kinds of audiences: your coworkers, your family and friends, your clients, your fans, and many other kinds of readers. This chapter will give you some tips and advice for how to take audience into account during your writing process.

Audience is a generic term. We don't always know the make-up of the audience receiving our work. And even when we think that a specific group will read our work (like certain professional groups, ethnic groups, religious groups, or peer groups), the members of that group won't all receive the work in the same way. No audience is so homogenous that one message will mean one thing to all of its members. For instance, think about yourself as a member of the group "college students." Not all college students think in the same way or share the same opinions. Just because something is written for an audience of "college students" doesn't mean every member of that group will agree, like, or even care about that message.

Still, recognizing the potential audience is important. Even if audiences aren't homogenous, we can still make some assumptions about audiences when we write. There are tools and strategies you can use so that an audience, whoever it is, will be interested, angered, encouraged, provoked, saddened, helped, or moved in some way. Your goal as a writer should always be to create a reaction in your audience. If your audience experiences no reaction to your work, then there was no reason for them to read or watch what you created.

What follows are some strategies for thinking about how to reach your audience. These are not the only strategies writers can engage, but you can use these as starting points for your work.

APPEAL TO CONVENTIONS, GENRES, EXPECTATIONS

Each profession has its own demands and interests. In the citation chapter, we noted that citation style is a professional interest. Using MLA for your citations, for instance, is an appeal to the professions that recognize MLA as the appropriate style. Other professions use different citations styles, such as APA. If you use APA style when writing for audience of social science professionals, for instance, you are recognizing that readers who study Psychology will understand your citations.

In that sense, citation style shows recognition of your audience.

There are other professional recognitions to consider as well. Some of these recognitions are formatting based: headings, subheadings, salutations, closings. Some of these recognitions are content specific, such as using the right terminology (people who work in real estate use one set of terms; people who work in medicine use another set of terms).

In the business world, a list of my professional qualifications and experience would be commonly called a "resumé." However, in universities and colleges, the resumé is referred to as a "curriculum vitae" (or CV for short). If I apply for a job as a professor at a university, my letter will use the phrase "curriculum vitae" and *not* "resumé." This is because my audience—other academics, in this case—does not use the term "resumé" on a regular basis.

Another way of saying this is that all writing is part of a specific **genre.** A genre is an audience-directed form of writing. The writing associated with a genre is carefully constructed so that the reader or viewer will recognize these conventions. Imagine yourself watching a movie where the main character finds herself all alone in a dark, creepy, old house filled with cobwebs. She moves slowly down the creaky floor toward a spooky closet door that is shut. She puts her hand on the old doorknob. The audience might instinctively shout,

"Don't open the door!" Why not? Because in the genre of horror movies, opening a closet door in a dark, creepy house often means trouble! The filmmakers know how to construct the conventions—darkness, old houses, cobwebs, the lone character, creaky floors—so that the audience recognizes the genre: **horror!** We expect something spooky to happen next because we know the genre's conventions.

So, too, does the romantic comedy have audience expectations. Think of a movie like *When Harry Met Sally* or *Pretty Woman*. The romantic comedy is a professional genre as well; it is a movie industry genre. An audience watching a romantic comedy typically has expectations:

Boy meets girl → Boy gets girl → Boy loses girl → Boy gets girl again

Again, there is variety to this organizational scheme, but it typically is the basis for a romantic comedy because *audiences* expect this structure.

All genres work with audience expectation. A school essay is also a genre: your teacher expects you to fulfill its conventional requirements. To do well on an essay, you need to know the teacher's (the audience's) expectations. In school, those conventions might include proper headings, numbered pages, works cited or bibliography, typewritten pages, and so on. Here are a few other examples of genres and audience expectation.

White papers can look very different, depending on the context in which they appear. Some white papers are very long, while some may be fairly short. Nevertheless, because the white paper is a professional writing genre, it does need to address some basic audience expectations. An audience reading a white paper typically expects the following:

- An introduction
- Background information
- Solution or solutions
- Conclusion

There can be variety to this organization, but overall, this is what is expected. Just like the audience of horror film genres has certain expectations, professional writing genres like white papers also have audience expectations.

A **white paper** is a good example of a **professional document,** one that comes with certain audience expectations. A white paper is a type of report that defines an issue, outlines a problem, or markets a product for a business or organization.

PLANET ARGON
MAKE IT HAPPEN

Responsive Design vs Mobile Site:

The Benefits and the Drawbacks

By Brian Middleton / May 2012

Responsive Design

Benefits

- Use the same code and content throughout
- Present a custom experience on a wide range of screen sizes

Drawbacks

- Different design variations need to be considered
- Some duplicated content may be necessary

Mobile Site

Benefits

- Easier, faster to implement and less expensive to launch
- Only the most essential information is presented in a streamlined interface

Drawbacks

- A separate silo of content is created that needs to be maintained
- Usually only targets one screen size
- Could result in higher maintenance costs over time

Determine Which Mobile Strategy Best Fits Your Website

When planning your mobile website strategy, you will want to ask yourself a few questions. Do you want your mobile site to operate separately from your current website? Would it be better, if you are starting from scratch, to build a site that could act as both a mobile and a desktop site? Can your current site be retrofitted to be more mobile friendly?

*Images not to scale

These are all important questions as you start down this road of mobile design. Your answers here will give you a good road map as to what kind of site you will be developing. If you are starting with an extensive existing site with a good deal of legacy code, you may want to go the route of creating a separate mobile site. If you are starting from scratch or have a smaller site, you may want to go the route of a responsive design. To help you through this decision, let's take a closer look at these different approaches and the pros and cons of each.

Learn More About Us Planet Argon has be designing and developing web applications since 2002 and active members of the Ruby on Rails ecosystem since 2004! Check us out at: planetargon.com

Figure 1. An example of white paper.[1]

The white paper is just one example of a professional writing genre. There are many, many genres and conventions associated with different kinds of writing. You will find it helpful to ask yourself, *"What are the audience expectations for the genre in which I am writing?"* What do typical readers of this genre expect you to do? Do they expect a certain kind of introduction? Do they expect anecdotes and personal stories? Do they expect the pages to be formatted in a certain way? Do they expect certain kinds of evidence and citations? Every type of writing fits in some type of professional genre scheme. Your task is to consider the expectations of that genre before you begin writing.

Appealing to your audience's expectations is one way to start thinking seriously about your audience. There are other types of appeals to consider as well.

APPEAL TO INTERESTS

Appealing to an audience's interests is another way to get your audience's attention. We can't always positively know what our audience is interested in, but we can make informed assumptions. By "informed assumptions," we mean that writers should go beyond stereotypes or common beliefs about groups.

Sometimes writers make the mistake of lumping whole categories of people together, just because they share demographic features: gender, age, ethnic background, political affiliation, birthplace, religion, etc. Not all women love shopping for shoes. Not all teenagers love playing video games. Not all Kentuckians love horses. Instead of relying on stereotypes, therefore, we can consider some other ideas for how to make informed assumptions about our audience's interests.

Consider this article in the *Dallas Observer,* **"Shock Top Sent Me $100 Worth of Swag and Two Awful Beers. That Sums Them Up Perfectly"** *(http://blogs.dallasobserver.com/cityofate/2013/04/shock_top_sent_me_100_worth_of.php).*

The author received a box of campfire swag (free merchandise) and two Shock Top beers because InBev (the brewery conglomerate who sent the products) hoped the author would write a positive review on his website. That review, in turn, would help sales. But there was just one problem. The author feels that the box of campfire-related products (Sterno can, stowaway pot, and items to make s'mores) is targeted not to a craft beer drinker (as the brewer intends) but to a stereotype. A craft beer drinker is someone who drinks beer from smaller breweries, which do not use artificial ingredients. Unfortunately, Anheuser-Busch, the brewery who makes Shock Top, uses artificial ingredients that craft beer drinkers find objectionable:

> *Craft beer is booming, taking a bigger slice of the beer market every year, and the foreign-owned corporations are scared shitless. But rather than improve their products by not using trash ingredients, they're spending money to play craft-beer dress-up, buying out breweries or creating their own pseudo-craft companies. This package of swag is such a brazen insult it almost seems like satire. It's more than $100 worth of marketing swag that reduces its badly missed target audience to a stereotype, and, almost an afterthought, a grand total of two beers. You couldn't ask for a more literal illustration of their priorities.*[2]

In this case, targeting doesn't work. Shock Top made an assumption about its audience's interest: *Craft beer drinkers will want these camping products and, thus, our beer.* In this case, though, this assumption was a misinformed one. As the author makes clear, the targeted audience of craft beer drinkers does not want cheap swag; they are more interested in improved products and better ingredients.

Facebook and other social media are great places to think about appeals to interest. Many people post to social media what they believe, what their political views are, which sports teams they follow, and what they do for fun. Users do so because they often believe they are appealing to their friends' interests: *"If I like X and Y, my friends probably do, too."* In this way, appeals to interest are based on an assumed commonality (the way Amazon will show you what it thinks you are interested in buying because of what other people, like you, bought).

Assuming others share your interests, though, can be risky. When you appeal to a supposed common belief, you can just as easily be appealing to what someone does not believe in. Sharing your interest so easily can backfire. Media consultant Arik Hanson once had a Facebook status update about this issue.

Figure 2. Advice from a social media expert about posting political opinions on Facebook.[3]

Merely sharing, then, is not without risks. Just because you believe in X or Y, doesn't mean someone else does as well. A client, as in Hanson's case, may abandon you if he or she discovers your public sharing and disagrees with you. Or it may cause the client to enter into a different relationship with you. Reactions are not entirely predictable. On the other hand, figuring out others' beliefs can help you later appeal to interest. Just make sure you know what the possibilities are when you share interests.

Because people are eager to express their interests, Facebook can target users with advertisements based on those expressed interests. On Facebook, we recognize that not only are we addressing an audience (the friends who read our updates), but that we are an audience for someone else (advertisers paying Facebook to be connected to you via your updates).

Beyond the question of what is or is not ethical about directing ads to Facebook users, this practice of targeting is a matter of audience identification. In the example of Facebook's targeting, advertisers are trying to make a connection between our status updates (our writing) and what interests us (as well as those people who are friends with us).

APPEAL TO STICKINESS

In their book *Made to Stick*, Chip Heath and Dan Heath describe this process of targeting ideas toward a specific audience. The sticky nature of a message, they write, is based on a few things:

- **Simplicity:** Keep things simple
- **Unexpectedness:** Generate interest and curiosity
- **Concreteness:** Use specific imagery
- **Credibility:** Demonstrate credentials
- **Emotions:** Make people feel something
- **Stories:** Perform a story[4]

For the Heaths, these six principles are often found in sticky ideas. Using six principles when targeting an audience may sound simple, but it's actually quite complex.

Advertisements recognize their potential audience (those who will buy the company's products) by often appealing to that audience's interests.

Apple's "Think Different" campaign is one such example.

Figure 3. According to Apple, Albert Einstein "think[s] different."[5] *(Image Courtesy of the Advertising Archives.)*

This ad appeals to its audience (young, creative, technology users) by saying

> *Albert Einstein was the type of person who could think differently.*

Apple sells products to help you think differently.

If you want to "think different," buy something by Apple.

Does the "Think Different" advertisement follow the Heaths' idea of being sticky?

- **Simplicity:** It uses one image.
- **Unexpectedness:** One doesn't usually associate Einstein or the other figures featured in the ads in the series with a computer (in fact, Einstein precedes the personal computer).
- **Concreteness:** Einstein and the other featured figures are recognizable (the person who lives next to you, for instance, would not work in this ad).
- **Credibility:** Einstein is a highly credible figure for his work in physics.
- **Emotion:** The figures tap into our emotional connections; we think these are smart or creative people.
- **Stories:** If Einstein was so creative and innovative without a Mac, can I be like him with a Mac?

Discussion Questions

1. What's the audience interest being appealed to in the "Think Different" advertisement?
2. **Levi's® "Go Forth" commercial** *(https://www.youtube.com/watch?v=FdW1C-jbCNxw)* also appeals to its audience (young jeans-wearers) by appealing to a series of tropes and ideas that might interest them. While the narration of the video is the nineteenth-century poet Walt Whitman, the audience for this video does not need to know that fact. Instead, the keywords of the poem,

   - Grow
   - Freedom
   - Enduring
   - Strong
   - Ample
   - Equal
   - Fair

   are meant to be appealing by telling a story young people will relate to and identify with. These terms are meant to create the feeling of having spirit, conviction, the desire to "go forth" and do good. Levi's wants its audience to make a connection between the idea and the jeans. Can you apply the Heaths' principles to this video as well? How?

APPEAL TO KNOWLEDGE

Specific audiences bring specific knowledge to their reading experiences. Audiences' backgrounds can greatly affect what they read and what they understand. Different readers will draw upon different kinds of *cultural knowledge* in order to understand what a text means. Knowing your audience often means knowing the kinds of cultural knowledge your audience has.

Tapping into an audience's cultural knowledge can be powerful. It can grant you the opportunity to share a message in a specific way or to make an argument more persuasive than if you used generic ideas or references as evidence.

One way to appeal to your audience's background or cultural knowledge is to use **allusions.** Allusions are references to what the writer believes will be understood as shared knowledge. We allude to people, time periods, historical moments, popular culture, famous sayings, and other information we expect the audience to already know.

On January 18, 2013, National Public Radio ran a piece on the usage of allusions or references in popular culture (Found here: **In a Fragmented Cultureverse, Can Pop References Still Pop?** *http://www.npr.org/2013/04/14/177204413/in-a-fragmented-cultureverse-can-pop-references-still-pop*).

The hosts of the show point out that many times, TV shows or performances reference texts that they don't expect their entire audiences to be familiar with. But the mere act of referencing, the NPR story shows, can still speak to larger issues the writers want to express.

How does a simple allusion help an audience to understand complex or large issues? It works by "anchoring" meaning in the text. Imagine for a moment that some complex issue under discussion is a boat floating on rough waters. Without any kind of anchor, that boat may well get carried away by the waves and wind. If there is an anchor, however, that boat will likely not get lost in the chaos. The same might be said about complex issues being debated. If there is an "anchor" to an outside reference, then your audience may better understand why this issue is relevant or interesting. By anchoring difficult topics to something closer to home, writers can help their readers better understand their meaning.

Let's take a look at a few examples of this how allusions appeal to a common background or cultural knowledge.

Allusions in Politics

Political campaigns, protests, and other acts involved with political issues often appeal to audiences by making allusions.

A popular **political allusion** that tapped into its audience's cultural knowledge involved signs carried by African American sanitation workers striking in Memphis in 1968. The signs were addressed to an audience of white employers, but also a larger American viewership watching the protests from home on their TVs.

Figure 4. Martin Luther King Memorial March for Union Justice and to End Racism, in Memphis, TN in 1968.[6] (© Builder Levy)

"I am a man" is an allusion to the U.S. Declaration of Independence: All men are created equal. By saying "I am a man," the striking workers are saying: Our Declaration of Independence says all men are created equal. You, our audience, know this. You also know that we are men. Then, treat us equally.

We also can call this type of allusion an **enthymeme.** An enthymeme is a type of reasoning in which a major or minor premise is hinted at, but not explicitly stated. Without all of the premises presented, the reader still is able to reach a conclusion.

For instance,

- **Major premise:** We hold these truths to be self evident, that all men are created equal.
- **Minor premise:** African American garbage workers are men.
- **Conclusion** (implied on sign): Treat me equally.

This allusion is enthymemic.

Another example is more recent. During Barack Obama's first campaign to be president, the Hope poster created by Shepard Fairey was widely disseminated at political rallies, in cities, online, and elsewhere.

Figure 5. The Hope poster, created by Shepard Fairey.[7]

Since the Hope poster's creation, many other visual editors have alluded back to this image in order to make a point. If you go to Google Images, for instance, and search for "Obama Hope Poster," you will find thousands of similar images, from R2D2 to Bob Hope to Steve Jobs to Alfred E. Newman to Big Bird. The creators of these images expect that you, the reader, will make the connection between the original poster and the newer image. You won't connect Obama's policies per se, but you will, it seems, connect the election's spirit of hope and renewal that many of Obama's supporters initially felt.

This connection may be satire (Alfred E. Newman, Big Bird). Or this connection may be about applying hope elsewhere (Steve Jobs).

Like the previous example, the usage of the Hope image as allusion depends on an audience reaching an implied conclusion based on the image's familiarity and previous association with Obama.

Allusions in Popular Culture

In **popular culture,** we often encounter allusions to other popular culture items. The writer or creator of the popular culture work (song, video, TV show, movie) will make an allusion with the expectation that the audience will recognize the reference and understand the overall meaning. In turn, the text itself, the author believes, will be stronger because of the allusion.

Because most of us listen to music and watch movies and TV shows, popular culture is an appropriate example regarding how to use allusions.

For instance, *The Simpsons* is a show with episodes full of allusions. Consider this brief clip of Bart strutting from season 6, episode 7, entitled **"Bart's Girlfriend"** *(https://www.youtube.com/watch?feature=player_embedded&v=KXtj_SdJMzM).*

The combination of strutting and Bee Gees' music is an allusion to the movie *Saturday Night Fever. The Simpsons* writers must have felt that audiences would understand the allusion and get the joke. The writers expected a significant portion of the show's audience to have seen *Saturday Night Fever* and to remember John Travolta strutting down the street in the film's opening scene. The writers did not pick an obscure film reference (though that might work for another show). But instead, they relied upon a reference to a film which has been widely viewed and whose opening scene is considered iconic by some, as well as representative of the intense sexuality of the 1970s (the time period of *Saturday Night Fever*).

When Bart struts, we make a connection to *Saturday Night Fever.* We understand the strut as John Travolta's strut. We make an emotional connection (Bart is like John Travolta!).

Film director Quentin Tarantino also uses allusions for similar purposes. He expects his audience to understand his movies' references to popular culture trivia, Kung Fu films of the 1960s and 1970s, and Blaxploitation movies. He also likes to name drop popular culture references. By doing so, he creates a relationship with his audience so that the films feel a part of everyday discourse.

Take a look at **this collage of Tarantino popular culture references** *(http://www.collegehumor.com/video/6860507/every-pop-culture-reference-from-tarantino-movies)* set up on a timeline from the 1900s to the present.

See how many references and allusions you recognize. By drawing on over a century of culture references, Tarantino shows us that we can always find a way to anchor a text for our audiences. A fight scene in *Kill Bill* is transformed, for instance, when O-Ren and Beatrix Kiddo exchange the parts of **"Silly rabbit, Trix are for kids"** *(https://www.youtube.com/watch?v=VI0mYThtLWg),* the reference anchors the scene. The reference to a 1970s cereal commercial appeals to audiences who grew up seeing it on television. The reference, in turn, transforms the tension of the scene into something slightly playful and funny.

If the scenes in question did not use all of these references, would the films still be as meaningful? If in *Pulp Fiction*, Jules didn't try and calm Yolanda down near the film's end by saying **"Be like Fonzi,"** would the scene be as powerful? Would it have the same meaning? Why not? What does the allusion to Fonzi do for the scene? How does it transform the robbery of the restaurant?

APPEAL TO PROVOCATION

While we sometimes feel that we have to appeal to our audience's needs, knowledge, or interests by being accommodating, we can also use the same knowledge to provoke. We don't always have to try to win over an audience. We don't always have to appeal to an audience. In fact, there are times where a writer needs to provoke her audience in order to:

- Get a point across.
- Engage in a debate.
- Push a reader's buttons.
- Start a conversation about a hot topic.
- Attract attention.
- Place the idea at the center of a conversation.
- Get the idea noticed.

In 2013, a high school senior named Suzy Lee Weis wrote an Op-Ed for the *Wall Street Journal* entitled **"To (All) the Colleges That Rejected Me"** *(http://www.wsj.com/articles/SB10001424127887324000704578390340064578654).* In the short essay, Weiss complains about not getting into college because she was "just being herself."[8] Instead of engaging in a number of extra-curricular activities, as many other classmates did in high school, Weis felt that she followed the popular advice to be herself. Being herself, it turned out, did not get her into college; her classmates did get into college. Once rejected, Weis felt that she had been treated unfairly for not doing what other classmates did in order to be accepted to college.

Whether an actual complaint or a joke, Weis's piece provoked enough people that it was covered in a number of other publications including ***Gawker*** *(http://gawker.com/5993140/attention-students-just-being-yourself-isnt-a-skill-that-should-earn-you-admission-to-college),* the ***Washington Post*** *(http://www.washingtonpost.com/local/education/pittsburgh-teen-bitterly-lashes-out-at-colleges-that-rejected-her/2013/04/02/dc60ce5e-9b9d-11e2-9bda-edd1a7fb557d_story.html),* **the *Today* show's website,** and elsewhere. A Google search for the Op-Ed brings up over 200,000 hits. The response to Weis's essay was very intense. By pushing the right buttons, she found her audience. In the comments sections of many of these online publications, readers complain that Weis is selfish, spoiled, privileged, and lacking any sense of what it takes to get into college.

Whether these critiques are true is not evident. But if we look at some of the topics Weis discusses—the things she feels she didn't do—we can see what provoked her audience so strongly:

- She mocks the SAT.
- She mocks charity.
- She mocks diversity.
- She mocks high school students who work part-time.
- She mocks tiger parents who encourage their children to have hobbies.
- She mocks high school students who have internships.

In other words, she finds the buttons to push. Weis mocks the categories typically used to determine whether or not someone should be admitted to college: high SAT scores, extracurricular activities, and highly respected hobbies. By mocking these categories, Weis made her audience uncomfortable. She made her audience angry. And the result was highly effective. She reached a very large audience by being provocative. Instead of only being read in the *Wall Street Journal* (a major feat by itself), she was read in many other online spaces.

This example doesn't mean a writer should always provoke. After all, your audience may find your provocations completely objectionable. You risk angering the very people you want to reach. If one goes too far (using racist, sexist, discriminatory, or insulting language, for example) one's ideas might reach a large audience, but the response could be detrimental (losing one's job, being ostracized, being accused of hate speech, etc.).

However, Weis's example does show that provocation can sometimes be productive, particularly when the goal is to reach a larger number of people. In Weis's case, she did not engage in hateful speech, but pushed many people to consider her ideas in a dramatic way. We might make the argument that Weis was successful in her provocation.

Here is another example of how writers sometimes appeal to provocation in their attempts to reach an audience. Roger Baylor is the owner of New Albanian Brewery in New Albany, Indiana, right outside of Louisville. Baylor runs a blog called ***The Potable Curmudgeon*** *(http://potablecurmudgeon.blogspot.com)*. Baylor, who is passionate about local food and local beer, can be quite provocative in his online writings.

In one blog post, **"Second Hand News Usually Is Mistaken"** *(http://potablecurmudgeon.blogspot.com/2013/03/second-hand-news-usually-is-mistaken.html),* Baylor recognizes the effects his writing may have on his audience.

> *Words I write are published alongside my name, because to me, anonymity is tantamount to cowardice. For those like me with strong views, there is an inescapable element of living and dying by the rhetorical sword, and I accept this condition of the engagement.*
>
> *Give and take is common, but every now and then, a complaint will be registered to the effect that someone, somewhere, has taken offense at words I've written. Actually I'm delighted with such feedback, and quite willing to discuss particulars, so long as we're reasonably clear about parameters.*[9]

Over a series of posts, for instance, Baylor has expressed anger at an Indiana liquor chain named Big Red Liquors. Accusing the chain of being a monopoly and of ejecting his brewery from a **Big Red sponsored event in 2003** *(http://potablecurmudgeon.blogspot.com/2012/04/nine-years-later-and-absolute-power.html)*, Baylor has critiqued Big Red several times and argued against the chain stocking beers from his brewery. **This post** *(http://potablecurmudgeon.blogspot.com/2013/03/big-red-liquors-pursues-its-monopoly-in.html)* begins with a specific provocation.

By posting a picture of the middle finger, Baylor is trying to:

- Provoke Big Red Liquors as potential audience.
- Let his other readers know how angry he is by featuring a provocative image.

In his recounting of the event *(http://potablecurmudgeon.blogspot.com/2012/04/nine-years-later-and-absolute-power.html)* that led to his conflict, with Big Red, Baylor describes the liquor chain's representatives in a provocative manner:

> *Our walking papers were verbally delivered to us by two members of the Big Red Liquors management team, John Glumb and Wade Shanower, a pair of wholly corporate, well-fed and utterly plain men who differed from the other polo shirts in attendance by the shared habit of spluttering ominously, brilliant*

white teeth clinched, blue neck veins bulging in a most unhealthy way and misshapen, contorted faces as red as their Big Red Liquors knits.[10]

Notice Baylor's usage of language:

- "Corporate, well-fed and utterly plain men"
- "Polo shirts"
- "Brilliant white teeth clenched"
- "Contorted, red faces"

The descriptions, entirely unflattering, are meant to have a provocative effect.

Figure 6. The label featured on one of Baylor's beers (and his blog). Image by Roger Baylor.

In addition, if you look at the figure above, you will see Baylor's own version of the Obama poster. Hope, in this case, is replaced with Hops and an image from one of the brewery's beer labels.

By openly critiquing a major Indiana retail outlet, Baylor is specifically using provocation as a way to address his audience. This provocation might:

- Position him and his brewery as independents in a market dominated by conglomerates and chains.
- Establish for his audience (those who visit his restaurant/brewery and buy his beers) an ethos as fiercely independent.

Whereas provocation helps Weis get her ideas to a large audience, provocation helps Baylor create a specific ethos. Both are productive usages of provocation.

APPEAL TO CLARITY

Most of what we've discussed so far deals with issues of content or structure. We are figuring out how to reach an audience by writing specific content or organizing material in familiar ways. Sometimes forgotten in discussions of audience is the importance of **proofreading.** Proofreading means going over your work so that you discover any typos, spelling errors, grammatical errors, or wrong information before an audience receives the work.

Why is that important? And what does proofreading have to do with audience?

Think about it this way. An audience reads your work. You don't want to lose credibility (ethos) over spelling or grammatical errors. Often, errors are the result of forgetting about audience. A typo not caught in proofreading can indicate that the writer is not paying enough attention to his or her own work.

Figure 7. Sometimes a simple misspelling can cost you customers.[11] (Photo courtesy of Rob Brink.)

Once your audience sees your error, they may think less of you as a source of information. Your readers may believe that a writer who is unable to fix a simple grammar error, or who misspells a major word in the first sentence, is not a writer capable of forming an important idea.

Let's imagine I'm applying for a job. I write a letter applying for a job, but I do not proofread it before sending it off. The letter misspells the addressee's name, has a run-on sentence in the first paragraph, has two more misspelled words in the first few sentences, and gets the company's name wrong. The audience for this letter—the person deciding on interviews—is not only likely to reject my application, but is also likely to stop reading early on.

Unfortunately, proofreading is more than just reading over your work. Because our eyes may not catch simple errors (we *think* we wrote the word or sentence correctly, but our eyes don't see what we really wrote), we often need someone else to read our work.

Always ask a roommate, friend, classmate, or someone else at the University to look over your work before you turn in it. At the very least, you will get an audience's impression. You can use that impression to catch errors or refocus certain parts of your work.

Ask your readers to mark any places that are confusing or difficult to understand. They may also be able to tell you whether they found themselves easily following your line of reasoning, or whether the organization needs some help. Asking for help with proofreading and feedback is not just something that beginning writers do. In fact, professional writers often ask many people to read over their work.

Exercise: Advertisements for Yourself

When we say writers "should know their audience," we mean that writers need to consider who may be reading or watching their works. But we also mean that writers need to understand how the words they use, the images they create, the stories they tell will forge connections with the audience in question.

Another guideline for reaching your audience might simply be: *forge a connection.* Much of what is discussed in this chapter is how to forge connections in different ways.

Practice some of the rhetorical tools described in this chapter by making an advertisement for yourself.

Using Chip and Dan Heath's principles of stickiness, as well as the other tools described in this chapter, turn yourself into an advertisement:

1. Think of the kind of story you wish to tell about yourself. (Examples: You as a great runner. You as a world-class sandwich maker. You as a wonderful roommate.)
2. Target it to a certain audience (perhaps your dorm, class, neighborhood, or friends).
3. Use a dominant image and/or some shared cultural allusions.
4. Consider what kind of *genre* your advertisement will be and what the expectations your audience might have for that genre are.
5. Consider whether or not you will use provocation in any way.

Student Writing Example: "Writing on the Wall" by Catherine A. Brereton

In the following award-winning essay, University of Kentucky undergraduate Catherine A. Brereton tells a personal and emotional story about violence against women. She draws on a number of different types of research to help make her point. As you read Brereton's essay, try to identify the techniques she uses to reach her readers. How did you react to her words and the stories she shares? What makes her writing so effective?

> *Nikki Burdine wants to be my facebook friend. Nikki Burdine is something of a minor celebrity; she's an occasional news anchor and frequent reporter for one of our local television channels. Just this morning she popped up on the morning news show, covering for an absent colleague; I watched her on the small screen hooked up to the elliptical machine at the gym as I struggled through half an ungodly-early-hour of going nowhere.*
>
> *I don't know why she wants to be my facebook friend. I'm in Starbucks when the request pings in on my phone screen; I managed to grab one of the few comfortable chairs and I'm alternating between research, writing, and knitting. Mostly the knitting wins; it's late spring and my motivation for research died along with the end of the semester.*
>
> ☙❧☙❧☙❧
>
> *Amongst other things, I'm researching violence against women, and Hanna Rosin, the author of the hard-backed book currently stuffed down the side of my chair and gathering crumbs from other people's cookies, chirpily tells me that crimes against women are lessening. She places, amongst the neatly cherry-picked statistics, a quote from a noted criminologist who firmly believes that women "might as well be living in Sweden, they're so safe."[1] So safe that one of the girls in one of my classes was raped less than a year ago by some guy she met in a bar. She wrote about it in a reflection piece for the class; our instructor—with the girl's permission—read out loud her description about being pinned down and penetrated while her housemates sat just yards away, on the other side of her closed bedroom door.*
>
> ☙❧☙❧☙❧

I accept Nikki Burdine's friend request. A few minutes later, a message flashes onto the screen. She's researching for a news report and thinks I might be able to help her. Is there some way she can meet me, that day, if possible? I type back that I'm in Starbucks, is that any use? She tells me that she can be there in half an hour, if I don't mind waiting. I ask her what the news report is about; I'm sure she's contacted me by mistake.

"I'd like to talk to you about Leigh Anne Kinder."

ઌ ઌ ઌ

Kelsey isn't feeling good. She has 926 facebook friends, and in her profile picture she's smiling out at them from beneath a black and white beanie hat. She's dyed her red hair black since I last saw her, and it looks like she's had her lip pierced, but the picture is grainy and I can't quite tell. She looks pale, though, and her smile is shallow, barely turned up at the corners. Last Saturday, she posted a status about the trial. It's seven weeks and one day away, and she's terrified.

Kelsey regularly goes to counseling. Sometimes, she posts about it, but usually it just remains unsaid, the only clue being her weekly headaches, near-constant insomnia, and a vague, general feeling of malaise which she communicates to her absent friends via the news feeds of facebook. I often think of posting some kind of reassuring comment on her facebook wall, but always hesitate to the point of cowardice. I belong to a different time in Kelsey's life, and watch now from the sidelines as she tries to put the pieces back together.

ઌ ઌ ઌ

We learned about Leigh Anne via facebook. It's confusing, at first; facebook's new security settings toss up all kinds of things in my news feed and it seems I'm getting notifications of people who have been posting on Leigh Anne's wall. I haven't spoken to Leigh Anne in person for maybe two years, not since she moved back to West Virginia to be with her not-quite husband and his family. He had struggled to find decently paying work here, and was going back, I was told, to the mines. They were going to live in a trailer up the road from his family; Leigh Anne would have more support and her not-quite husband would have a job. Things would calm down.

But she and her not-quite-husband split up not long after. They shared custody of their young daughter; Leigh Anne kept her eldest daughter from a different father. She emailed once or twice to tell me how she was doing. Her ex-not-quite-husband had turned into quite a different man, given to temper and loud voice. She couldn't bear it, and they parted. She met someone new, new to her friends at least, an old high-school sweetheart, and she was infatuated. They were married within the month.

ଔ൬ଓ൬ଔ൬

The girl in my class was raped by a man she met in a bar. She complains in class, with as much bravado as she can muster, of the police response. She'd been drinking; that alone should constitute what happened as non-consensual, but they question her relentlessly as if maybe she'd said yes in some kind of drunken haze and had simply forgotten about it. They ask her about other sexual encounters; did she make a habit of picking up men in bars and taking them back to her apartment? She tells them that he had offered to walk her home; he seemed like a nice guy. No, she didn't know his full name. She has bruises on the inside of her thighs. They ignore the bruises and ask her again about her last boyfriend.

ଔ൬ଓ൬ଔ൬

It's almost 10 p.m. when the text alert sounds. A female colleague of my partner has been arrested. The woman has been accused of slashing another woman's face to ribbons on a Tuesday afternoon outside a dress shop in our local mall. It's been all over the nightly news, which we never watch, but tonight we turn it on to find out more. The victim's face was criss-crossed with what police believe was some kind of razor blade, and she's lucky to be alive—one centimeter lower and her jugular would have been severed. The news report tells us that the victim half-knew her attacker, that she had dated her attacker's ex-boyfriend. The attacker apparently took a friend with her to pin down her victim and hold her still while she carved up her face.

ଔ൬ଓ൬ଔ൬

Nikki Burdine arrives. She's petite, well-dressed, and incredibly perky, with a bright smile and perfectly white, perfectly straight teeth. Despite the humidity, her hair is immaculately coiffed and falls in rich waves onto her tiny shoulders. She's wearing stiletto heels and a short skirt, and she has a healthy, perpetual tan. As she clicks across the tiled floor, the men huddled in the comfy chairs next to mine stop their conversation to look at her. She scans the coffee shop and settles on me as the most likely person to be me, before introducing herself with a confidently outstretched hand, and a light, made-for-television voice that matches her appearance.

When she asks me about Leigh Anne, her voice slips into a practiced, pacifying tone, the same tone I adopt when I'm soothing an upset child. My voice lowers in unconscious mimicry.

"Did you know Leigh Anne's husband?" she says. "Did you ever meet him?"

I answer in the negative. The only thing I knew about Leigh Anne's new husband—and now, I suppose, technically her ex-husband, is that they once were high school sweethearts. She said on her facebook wall—or maybe it was in an email—that he made her smile. There's a photo of the two of them on her facebook page. It was taken the October they met, not terribly long after she moved away. He's sat on a dirt bike, and is pulling her in close, with his arm around her waist. The sun is shining directly at them; they're both squinting slightly, and one of his eyes is closed, as if he's winking at the camera. She has her hand protectively on his thigh, and her dark blonde hair is pinned back from her face by a pair of sunglasses. The first comment on the photo calls them an "adorable cute couple." The second comment says: "Its fuckin sad to see how he fuckin killed her."

ଔ ଔ ଔ

Hanna Rosin believes that feminists are "irritated" by the latest statistics from the White House, statistics which claim that "women today are far less likely to get murdered, raped, assaulted, or robbed."[2] *She writes of women-turned-poisoners, women who plan and plot and scheme "unprovoked, premeditated"*[3] *attacks on their husbands and boyfriends, and girls who shatter the windows of drive-through fast food restaurants because they can't get their preferred junk food fix exactly when they want it. Nicole Bewley, who has been accused of slashing Nicole Kyaw's face in a busy mall, purportedly planned her attack to such a degree that she took a friend with her. There is nothing opportunistic about having a friend hold down your intended victim so that you can disfigure her for life. Hannah Rosin states that "what looks like warped logic in one context can look like empowerment in another."*[4] *I look at Nicole Bewley's mugshot on our local news-station's webpage and wonder if this is the face of the empowered woman.*

ଔ ଔ ଔ

Kelsey has been enrolled at five different schools in three different states in the last eight months. She's lived—or so I gather from a careful perusal of her facebook—in five different houses with five different families, and she's been in foster care just the once. The couple that she lives with now has permanent guardianship, and Kelsey seems happy; she calls Laura her second mom and wonders what she would do without her. But Kelsey's half-sister, possibly her nearest blood relative, no longer lives in the same state. Her facebook wall hints that she's only seen her sister for maybe three or four days in the last eight months.

Kelsey has 58 profile pictures. After three of her in the black and white beanie is one of her new tattoo. It's on the outside of her calf, and takes up most of her lower leg. She has a large anchor, wrapped around a rope-edged heart. In the center of the heart, there's a lighthouse, and waves crashing onto a beach. Below the rope-edged heart, the words "the sun embraces the darkest shadow" wind their way around the base of the anchor. There's a date tattooed at the very bottom, and at the very top, above the anchor, the words "In Loving Memory Mom & Mawmaw" are indelibly marked on Kelsey's skin. She says it gives her a little more closure.

ઉ૪ઉ૪ઉ૪

Leigh Anne's mom, Gloria Sue, had been staying with her for a while. Leigh Anne's dad was terminally ill, and Leigh Anne needed both the emotional support and practical help of her mom. Gloria Sue was in the house when Leigh Anne's new husband slit his new wife's throat. News reports say that Gloria Sue hadn't seen her daughter for two days; Leigh Anne's new husband told her she was sick with the flu and didn't want to be disturbed. Gloria Sue checked on her anyway, only to find her daughter's lifeless body lying in an upstairs room. Leigh Anne's new husband slit Gloria Sue's throat too, and left both the bodies bleeding out onto the upstairs carpet while he (allegedly) raped Leigh Anne's oldest daughter, his step-daughter. His trial is in seven weeks and one day.

ઉ૪ઉ૪ઉ૪

Nikki Burdine wants to film an interview with me. We go outside to the parking lot where her cameraman is waiting, and in front of the coffee shop and a chi-chi cupcake store she asks me the same questions she has already asked me. I try to be articulate, but the sun is in my eyes and I'm squinting. The cameraman stops the interview and asks me to turn in a different direction, but then the cupcake shop is in full view and somehow that doesn't seem appropriate. Eventually, we settle on a position where, over Nikki Burdine's shoulder, I'm watching people pop in and out of the shop to buy their chi-chi cupcakes, but only the traffic whizzing past on the busy road behind me can be seen by the audience who will watch my interview on the six o' clock news.

☙❧☙❧☙❧

My partner calls me from work the day after her colleague was arrested. She tells me that Nicole Bewley no longer works for the company and that her electronic access card—the one that grants access to the private offices—has been revoked. She doesn't know if she is in custody or out on bail, but the offices are filled with quiet whispers about the colleague who slashed a woman's face; everyone wonders if they saw signs of inherent violence, or if maybe they said or did something untoward that might have justified some kind of violent attack. I believe Nikki Burdine was the reporter that first broke the story.

☙❧☙❧☙❧

Timothy Parsons (allegedly) showed Kelsey the dead bodies of her mother and grandmother before he (allegedly) raped her. (Allegedly), he told her he would kill her too if she didn't have sex with him. After he (allegedly) raped Kelsey, he left the house leaving behind two dead bodies and an (allegedly) raped teenage girl. It was 24 hours before Kelsey was able to free herself from the ties that had been used to restrain her during her (alleged) rape. Timothy Parsons was found the following day and arraigned without bail.

☙❧☙❧☙❧

Kelsey doesn't talk about the rape, at least not on facebook. Those details only came out in the subsequent news reports. When she talks about the events of last year, she talks of how much she misses her mom and her mawmaw, how much she wishes they were still here to help her, and how she'll make them proud of her. When she talks about the upcoming trial, she talks about justice for them, and "even" for herself, as if she's decided that being tied up and raped is somehow less noteworthy than death. She has an album on her facebook page of photographs from the funerals, filled with pictures of the shiny white coffins containing the bodies of her mom and mawmaw, and close-ups of the pink and purple floral displays, elaborate sprays of roses and carnations that dripped from the top of the coffins to the ground. The funerals took place on a sunny, green day. The trees in the background were heavy with foliage and the grass still retained the neon verdancy of spring, unmarked as yet by the brutal summer sun. The photos show family members dressed in bright colors, smiling at the camera, squinting when the sun hit their eyes. It looks like almost any other happy family event, apart from the two shiny white coffins in the background.

ઉ ઠ ઉ ઠ ઉ ઠ

After we've finished the interview, Nikki Burdine hangs around for a little while; her cameraman needs a cigarette and she needs a cold drink. She orders some kind of iced tea from Starbucks, which they mess up, so she waits with me while they make her another. Curious about my British accent, she asks what brought me to Kentucky, and we talk for a little while about immigration rights. Neither of us mention Leigh Anne. Her iced tea is finally ready, the cameraman has finished his cigarette, and she has to head back to the studio to edit the piece for that evening's news. When I watch the report that night all I can focus on is my large, pink face, like a disembodied head floating above Richmond Road as the traffic flies past and the people I know I can see over Nikki Burdine's shoulder buying their expensive cupcakes. My 16 year old daughter, once a good friend of Kelsey's, makes fun of my voice, which sounds plummy and pretentious alongside Nikki Burdine's soft Kentucky drawl. We don't talk about what happened to Leigh Anne and Gloria Sue. We don't talk about what happened to Kelsey.

That summer, I sit in Starbucks every morning. Instead of studying I knit Kelsey a shawl. It's a gesture that brings only me comfort. It's pink, and soft, but pink, I now realize, isn't really her color, and a shawl isn't really her thing. I knit it anyway, and send it the mail with a note telling her how sorry we all are, and that the shawl is the closest thing I can manage to a hug. I wonder if she ever got the shawl; she never mentions it and I don't want to ask.

ଋଌଓଌଋଌ

Nikki Burdine moves on to new stories: such is the nature of her job. Her news station picks up the story of woman with the slashed face, and then, doubtless, the story of someone else. There's always a new story, always another woman who has been murdered, raped, assaulted, or robbed, and the stories of Leigh Anne and Gloria Sue, of Kelsey, the girl in my class, and the woman with the slashed face, are quickly subsumed beneath more stories of more women, quickly forgotten, quickly reduced to statistics.

Those statistics report "plummeting rates of completed rape, assaults, attempts, and threats."[5] Criminologists report that women are "a lot harder to victimize," claiming that "people don't admit these trends because there is a lot of discomfort ... about girls succeeding so well ... while boys are on a destructive decline."[6] I think of the women I know who, in the last year, have been murdered, raped, and assaulted, and ponder the meaninglessness of this statistical decline.

ଋଌଓଌଋଌ

Leigh Anne's facebook page is still active. Every so often, my news feed will show that she has received a horoscope or inspirational quote. Her wall is peppered with daily Bible verses and images of her with "R.I.P" photo-shopped in alongside clip-art red roses and hearts. Sometimes, facebook suggests that I accept Gloria Sue's friend request and tells me that we already have two mutual friends; one of these is Leigh Anne, the other is Kelsey. All three of them defy the statistics.

NOTES

1 Hanna Rosin. *The End of Men and the Rise of Women.* 2012. Riverhead Books. Pg 17.

2 Ibid. Pg 182.

3 Ibid. Pg 178.

4 Ibid. Pg 182.

5 Ibid.

6 Ibid. Pg 183.

VIDEO RESOURCES

Stella Parks on Writing and Audiences

Stella Parks is a pastry chef (formerly the head pastry chef at Table 310 in Lexington), a blogger, and a cookbook author. Stella is a native of Louisville, but she has lived all over the world. As part of her daily work as a pastry chef and blogger, Stella uses rhetorical theory and writing in a number of ways. In this interview clip, she describes how she thinks about audience in her writing.

(https://vimeo.com/103938664)

Griffin VanMeter Describes How Social Media and Audiences Can Help Spread a Media Campaign

Griffin VanMeter is co-owner of the branding firm **Bullhorn Creative** *(http://bullhorncreative.com),* as well as the creative director of Kentucky for Kentucky *(www.kentuckyforkentucky.com)* Griffin grew up in Lexington, though he has also lived and worked around the country. As part of his creative work, Griffin uses writing and rhetorical theory in many ways. In this interview clip, he describes how audiences and social media helped spread the wildly popular "Kentucky Kicks Ass" video.

(https://vimeo.com/104015857)

You can also watch the "Kentucky Kicks Ass" video:

(https://vimeo.com/55873402)

NOTES

1 *Planet Argon*, last modified 2014, http://planetargon.com.

2 Jess Hughey, "Shock Top Sent Me $100 Worth of Swag and Two Awful Beers. That Sums Them Up Perfectly," *Dallas Observer*, April 4, 2013.

3 Arik Hanson, "Should PR Folks be Sharing Their Political Viewpoints on Facebook?" *Communication Conversations*, last modified 2014, http://www.arikhanson.com/2012/11/06/ should-pr-folks-be-sharing-their-political-viewpoints-on-facebook/.

4 Chip Heath and Dan Heath, *Made to Stick: Why Some Ideas Survive and Others Die* (New York: Random House, 2007), 16–17.

5 "Albert Einstein Think Different Apple Advertising," *Tomorrow Started*, last modified October 4, 2011, http://www.tomorrowstarted.com/2011/10/dumbing-down-a-nation-when-did-being-stupid-become-cool/.html/albert-einstein-think-different-apple-advertising-2.

6 *Campus Circle*, last modified 2014, http://www.campuscircle.com.

7 "Barack Obama 'Hope' poster," *Wikipedia*, last modified May 9, 2014, http://en.wikipedia. org/wiki/Barack_Obama_%22Hope%22_poster.

8 Suzy Lee Weis, "To (All) the Colleges That Rejected Me" *Wall Street Journal*, March 29, 2013.

9 Roger Baylor, "Second Hand News Usually Is Mistaken," *The Potable Curmudgeon* (blog), March 25, 2013, http://potablecurmudgeon.blogspot.com/2013/03/second-hand-news-usually-is-mistaken.html.

10 Baylor, "Nine years later, and absolute power still corrupts Big Red Liquors absolutely," *The Potable Curmudgeon* (blog), April 20, 2012, http://potablecurmudgeon.blogspot. com/2012/04/nine-years-later-and-absolute-power.html.

11 Robert Brink, *RobBrink*, accessed June 27, 2014, http://www.robbrink.com.

Strengthening Your Writing Process

Chapter 4

Let's say you've just been handed a writing assignment. You are eager to get started on your project and jump into writing! But, occasionally, students find this stage a little intimidating. How exactly should you start? How do you know if you're making a strong start? How do you know when you're done?

All of these questions are about the **writing process.** We call it a process because good writing does not simply pour out of writers already neat, perfect, and ideal. Professional writers hardly ever sit down and craft a finished piece of writing in one attempt. They begin with ideas, sketches, notes, and flashes of arguments. Then they work on writing, steadily, for a period of time. During this time, they might move around paragraphs or sentences they wrote the day before. They might expand on some ideas and make notes to revisit other ideas later.

Discuss Your Writing Process

What is your normal writing process like? What do you normally do after you get a writing assignment? Do you begin right away or do you wait a few days (maybe too many)? Do you usually write entire essays in one sitting, or do you give yourself some time to revise and rethink?

Which of these habits do you want to keep and which do you want to change? Why?

Once writers have whole drafts, the work of revision begins. Every writer revises a little differently. Some people like to read what they've written and revise as they re-read. Some people like to read their drafts without planning to revise right away. They simply get a sense of what they've done. Other writers like to get feedback from other readers before beginning to revise.

This chapter offers some suggestions for making your own writing process stronger, easier, and more rewarding. The suggestions are real tips that are taken directly from professional writers. Try them out next time you are faced with a new project.

STARTING FROM SCRATCH

Reading and Making Notes

One of the first things that professional writers do when facing a writing project is to *read*. They might read books and articles about the subject that they're studying, or they might read primary research that relates to their topic. No matter what they read, most professional writers find that they have to be familiar and invested in the conversations that are taking place about this subject. This is because it is really difficult, if not impossible, to simply come up with a brilliant idea off the top of your head. This is just not how our brains work. Everyone, even professional writers, have to listen to the voices around them in order to come up with something that is interesting and worthwhile to write.

When reading these things, these same writers will also make notes. It might be something as simple as a summary or a direct quotation with a page number of whatever they've read. But it might also be some ideas that were sparked during the reading. For example, I usually have a notepad with me, and I will write down page numbers and ideas that come to me as I read things that I want to respond to.

The advantage of reading and making notes is that you can go back to these notes during the "real" writing stage, and you can remember whatever it is that you want to respond to. Ultimately, it will make your life so much easier when you begin by taking notes and reading.

Brainstorming

Usually, after reading and making notes, the next thing that professional writers do is some kind of brainstorming activity. You have probably tried out some of these brainstorming activities in high school. These activities, like spider webs and free writing, are pretty typical ways of getting some ideas down on paper. Try out some of these exercises to see if they help you to form ideas.

Free writing is a very easy and quite productive brainstorming exercise. Give yourself a certain period of time, like five minutes or ten minutes. Don't make it too much longer because you can easily find yourself getting exhausted or tired of this activity. Sometimes, shorter periods of time are the best for getting some creativity going. Take out a pen and paper or sit at your computer and start writing for that set period of time. Don't stop at any time until your timer goes off. Keep writing. You can start by simply making the first sentence about your topic. Then write whatever comes to mind. It might seem to be directly related or it might not.

Don't edit yourself at this point. Whatever comes to mind, simply write it down. Once you've reached the time limit, *then* you can go back and reread what you've already written. Are there any ideas that came out of this session? Sometimes, what might seem like a crazy or unrelated idea is actually some part of your brain trying to tell you something. This is why you don't want to edit yourself during brainstorming sessions. If you found an idea that you really like, then you might start another short brainstorming session that more directly focuses on that new idea.

Talking

Professional writers also get ideas about what they want to write by simply talking to other people. This might mean talking to other professional writers, but it might also just be talking to friends or family. Sometimes you can get perspective on a subject by asking your friends or family what did they think about an issue or the subject. Talking out loud actually uses a different part of your brain than thinking internally. Some writers find it helpful to simply talk to themselves out loud, talking themselves through the idea and all of the issues in order to activate that different part of the brain. It might sound a little nuts, and maybe more than a little embarrassing, to talk out loud to yourself. For this reason you might find a quiet place to try this out. But once you feel safe and private, you may find that this process jumpstarts some ideas that simply were not flowing when you were thinking about the subject internally or quietly.

Moving

Just as talking out loud uses different parts of your brain, thereby jumpstarting some different perspective on your topic, many professional writers will actually move in order to achieve the same results. Many writers have talked about going for walks or runs while thinking about their subject. This is no accident. Physical movement also requires us to draw upon different brain functions, and for this reason you may find yourself thinking about your subject in new and creative ways when doing something as simple as going for a walk or run or even a bike ride. Sometimes, after I run, I come home and open up my laptop in order to write down all the ideas I've had during my previous hour's run.

Cubing

Some writers like to brainstorm by practicing something called "cubing." Imagine that you're looking at a six-sided cube, each side with a different activity on it. As you cube, you will respond to all six activities while thinking about your topic.

Get some paper and a pen or sit at your computer, and set a timer for three minutes. For three full minutes, write as much as you can in response to the first command. When you reach your time limit, start the timer over and do the second command. Keep repeating until you've finished responding to all six.

- Describe your topic
- Associate your topic
- Apply your topic
- Compare your topic
- Analyze your topic
- Argue for and against your topic

Making a Plan

After writers brainstorm their ideas, the next thing they usually do is make some kind of a plan. Sometimes these plans can look like formal outlines, but sometimes they look a little bit less organized. You might try a number of these different planning methods and see which one works best for you.

Outlines

You're probably very familiar with formal outlines. Outlining can be helpful because they help you to visually organize your ideas and arrange the argument so that as you write, you can see how your ideas link together.

Outlines can be divided into major sections and subsections, and some writers even like to annotate their outline so that they can see exactly where they plan to go. It can be easy to feel bound by a formal outline, but in reality, professional writers often revise their outline as they write. You might think of an outline as a kind of guide that changes as you write and think about your topic. Use your outline as an organizing principle and not something that you must follow to the letter of the law.

If you are the kind of person who feels too constrained by formal outlines, you might think more in terms of a *plan*. A plan is still an organizing guide that will help you to visualize how your argument unfolds, but it does not have the level of formality that an outline has. I often use plans when I'm writing, and I love to use headings and subheadings so that I can imagine how my argument might look when I finish. But, once again, a plan can also change and be updated as you write.

Proposals

You may also find that writing a proposal can be incredibly useful for planning your argument. In fact, you may be required to write a proposal before beginning your paper. Proposals are more specific than outlines because they also help you to imagine *who* your audience is, *why* you are writing, and *what* types of arguments you may want to make. Many writers find that the act of writing a proposal can help them to focus their attention and their arguments much more clearly than an outline does.

There is not one right way to write a proposal. You may find that you are given a set of directions that contain the type of information your audience will want in your proposal. For example, if you are writing a proposal for a community grant, the organization giving the grant may already have a list of requirements that must be contained in the proposal. Likewise, when you are writing a paper for class, your instructor may give you a list of questions and requirements that must appear in the proposal.

In general, you will need sections in your proposal that accomplish the following:

Define the Purpose

In the section on purpose you will need to explain to your readers what your project is going to do. What is the point of your project? What are your goals? What is it that you are aiming to accomplish in this project? Remember that your readers may not know anything about your project. They need to understand that there is some kind of exigence for what you are about to argue or do in your project. Think of readers as friendly, but perhaps skeptical and, at the very least, overworked. Why should they sit up and take notice of *your* project out of all the other ones that they may be also looking at?

Provide Background

This section helps your readers to understand what the current status quo is. This is your chance to persuade your readers that there is a *need* for your argument, your project, or your product. If you are writing about a subject that your readers may not be familiar with, this is also the place where you would give them enough information so that they can follow along with your proposal.

Describe Your Objectives

Your objectives are the ultimate goals that you are hoping to accomplish in your project. In this section, you will help readers to understand and visualize exactly what this project aims to achieve at its conclusion. You want to be very specific in this section so that your readers understand what your aims are and what you want to achieve. If you seem vague, confused, or unsure about your own directions, then your readers are unlikely to be persuaded that you can actually achieve this. Writing a proposal also allows you to demonstrate your own ethos to your readers.

Describe Your Sources or Research

You will want to thoroughly describe what types of research you will draw upon during your project. Explain why these sources are most appropriate to your goals and objectives. For example, if you plan to interview certain people, explain why those people are appropriate. Likewise, if you have sources you have found online, explain why those sources are credible, authoritative, and invaluable to your topic.

Exercise: Pitch

A proposal is also sometimes called a "pitch." You've probably heard some kind of a pitch before, even if it's just a basic sales pitch. When someone delivers a pitch, she or he is trying to persuade the audience to do, buy, or think something. The best pitches tell audiences:

- *Why* they need this product
- *What* this product does and what makes it so great
- *How* to use this product

Imagine you have just sixty seconds to pitch a brand new product that you've invented. You want your audience, who happen to be a group of wealthy investors, to invest one million dollars into your new company. Together with a partner, come up with your dream product and then outline your pitch. Be sure to include all the important elements of a pitch or proposal. Deliver the speech in front of your class!

Student Writing Example: Proposal Prepared for Habitat for Humanity by Kyle Taylor, Anthony Hendren, Matthew Ewadinger, Joelle Hill, Derek Woodruff

In the following proposal, University of Kentucky students Kyle Taylor, Anthony Hendren, Matthew Ewadinger, Joelle Hill, Derek Woodruff created a proposal for operational improvements at the charity organization Habitat for Humanity. The group's proposal is written with a professional audience in mind, and they hope to persuade Habitat for Humanity to adopt their recommendations. As you read, notice the different sections that the group includes in their proposal. How do they organize their background information, and how do they explain their ideas in a way that readers will easily and quickly understand? How does the group establish ethos in their proposal?

Habitat for Humanity: Proposed Operational Improvements

Prepared for

Habitat for Humanity
Operational Headquarters
121 Habitat Street
Americus, GA 31709-348
USA

Prepared by

Habitat for Humanity Analysis Group
Kyle Taylor, Team Leader
Anthony Hendren
Matthew Ewadinger
Joelle Hill
Derek Woodruff
WRD 203-006
University of Kentucky
Lexington, KY 40506

07 December 2012

7 December 2012

To: Habitat for Humanity Administration

From: Habitat for Humanity Analysis Group

Subject: Proposals to resolve potential issues in organization's charitable work.

This report was written with the purpose of finding areas of improvement within your organization. We were hired by Habitat for Humanity in their attempt to continually evolve with time so that they can help as many people as possible. Our proposals are not intended to suggest anything negative about your organization, and it is understood that you will not agree with all of our ideas. Our intention is to present ideas that may catch your attention and make your work more efficient.

Initial research showed that there were many perceived issues that we could discuss. We have found various news articles related to Habitat homes that had breakdowns after they were built, which led us to our first topic. Further research showed that there was evidence to prove that Habitat homes could possibly decrease property values, although we realize your organization has fought these claims. This research validates and supports our second topic. Finally, we found articles that claimed that your organization does not need to continue to build homes in this current housing crisis. We took this under consideration as our third topic. While other issues were discussed among the team, these were the issues we believed we could effectively resolve. We have chosen to focus on your work within the United States because it would be difficult to analyze all of your international work and compare it to work within our country.

From all of us at Habitat for Humanity Analysis Group, we thank you for taking the time to read our report on proposed operational changes.

Executive Summary

This report analyzes problems that occur in the Habitat for Humanity homes during and after their construction. There have been instances where recipients of homes have had issues in the years after they have moved in. We have also analyzed a few potential changes the organization can make that will help them save money and redistribute their funds to other causes.

The first issue we discussed was issues that occur in homes due to the use of unskilled labor. Research shows that the majority of Habitat construction sites do not have much tool training or safety training. Our research also shows that the biggest issues that occur are often with the planning and permitting phase. We have suggested that the construction department be split into an administration department for obtaining permits, and a separate department that focuses solely on the actual construction.

Our second issue involved the question of whether or not Habitat homes devalue neighboring property values. There is evidence to support this idea including an interview we administered with a neighbor of a Habitat home. These homes are usually built in low-income areas because the recipients are low-income families and it is best suited for them. Moving these homes to wealthier areas is not really an option due to opposition from people in these neighborhoods. We propose the utilization of a program known as the Inclusionary Zoning Housing Program (IZHP), which will be outlined in more detail in the body of this report.

Our final proposal was that Habitat for Humanity should consider a plan for repairing homes rather than building new homes. The recent housing crisis has left many homes uninhabited and the need for new homes has diminished. If Habitat could formulate a program that repairs existing homes and moves families into them they could possibly save money and help even more families.

We believe that our proposals can strengthen your organization, as it will allow you to help even more people as well as putting those people in a better position to succeed. We have done thorough research, performed multiple interviews, and administered a survey that supports our hypotheses. Habitat for Humanity is an outstanding organization and its effort to keep improving has been one of its most admirable qualities.

Table of Contents

List of Illustrations

Introduction

Habitat for Humanity (HFH) is one of the most influential charities in the world. Since HFH's creation in 1976, over 500,000 families have been served, 80,000 of which were in 2011. Unfortunately, there will always be people in need of assistance with finding a home. Our analysis group applauds the work that your organization has done and will continue to do in the future. The model that you currently use has proven to be very successful allowing you to help families continuously for over thirty years. We believe that with our proposals your organization can reach even more people. We also have done thorough research on potential issues that arise in the work that your organization performs. We have discovered some easily instituted solutions to these issues. The issues we will discuss are:

- Construction mistakes due to unskilled labor.
- Decrease in neighborhood property values due to a Habitat home.
- Moving toward repairing existing homes rather than building new homes.

Our proposals are not meant as an attack on your organization's policies. We believe that if the changes in our proposal are made, your organization can continue to grow and influence even more lives. The solutions proposed in this report involve plans to save money by repairing existing housing rather than building new homes. If money can be saved through this concept, then that money can be put toward helping even more families, as well as being invested in more qualified labor and more reliable materials.

Unskilled Labor

In order to help improve this area of HFH's foundation, we decided to research all the issues that arise from using unskilled labor and formulate some ideas in order to increase the effectiveness of utilizing volunteers and the overall success of the organization. When it comes down to the issue of unskilled labor, the first thing people wonder is whether or not unskilled labor possesses the skill needed to build a house. The answer to that can be yes and no. Unskilled volunteers are a major contributor to the overall process, but it could not happen without the experience and leadership of HFH's construction team. These members include a Construction Director, Project Manager, Coordinator, and Supervisors who all do an intricate part of the process from obtaining supplies, receiving grants, coordinating contractors, scheduling inspections, and lining up volunteers. Without the HFH construction team, HFH would not be able to function as successfully as they do and the unskilled volunteers would not be able to be such an intricate and essential part of the process. What makes up such a vital part of the process also contributes to some of the problems.

Volunteer Scheduling and Reliability

During our research of unskilled labor we realized through our interview with Kelsey Giauque, the Construction Coordinator for HFH for the Lexington chapter, that one of the biggest problems when it comes to the construction process, is the multitude of delays that result from many different variables. Variables, like the weather and contractors, cannot be helped and will always have to be worked around. If a contractor doesn't show up or takes longer than expected, Kelsey mentioned, then it backs up other contractors who have to rearrange their schedules in order to attempt to keep HFH's house construction on track. However, we want to focus on creating solutions to problems that we can make more efficient. Kelsey also mentioned that these delays include a mismatch of the amount of volunteers available to the amount of volunteers needed to complete specific tasks. For example, the Project Manager decides that Friday they will start framing, but not enough volunteers have shown up and so they now get behind or have to delay the framing altogether. This brings to light the issue with volunteer scheduling and reliability.

The problem with the scheduling is that it lacks the coordination involved in scheduling volunteers to tasks. It's too much to piece together for one or two people. They have to take into account the number of active construction sites, the number of volunteers, the types of tasks those volunteers feel comfortable doing, and the amount of volunteers needed to complete the tasks asked of them. On top of that, volunteers sometimes do not show up which can cause the tasks to be more difficult or delayed. Our proposal to combat this inefficiency and lack of coordination is to create a website that is user friendly and can be accessed by both HFH construction teams and by volunteers wanting to participate in HFH's great program. We hope that in creating this website, construction and volunteer coordinators can identify early enough what tasks are going to be completed where and how many people are needed there. After that is done, the website can allow volunteers to sign up for specific job sites where they feel their skills are needed most. The website will also have the technology of smart emailing, which is capable of doing two important and useful tasks. It can send out emails to past participants who match the upcoming tasks in an effort to fill up the quota of volunteers needed and it can send out emails to volunteers who are already signed up to inform them of a cancellation. With this smart online programming it becomes easier for both HFH and volunteers to plan and participate. We hope that this user-friendly site increases the amount of volunteers and return volunteers by developing a stronger relationship between HFH and HFH volunteers.

Safety Issues

When it comes to unskilled labor and tools, there is an increased potential for accident and injury. There is a common understanding that volunteers have some sort of construction or carpentry experience and some sort of safety sense prior to volunteering, however that is usually not the case. According to Figure 1 (as seen on the next page) in reference to our HFH survey, approximately 80% of participants said they had no construction experience prior to volunteering for HFH. That means at least 80% of 1st time volunteers are unaware of the safety hazards that are present in a construction environment. HFH contracts a good deal of their work out and so it takes away from a lot of the hazards that may normally be present during the construction process, but it is impossible to get rid of them all. You will always have potentially dangerous materials and objects lying around from tools, to uneven ground, and to raw materials. It would be nice if everyone was safety conscious, but we know this just isn't realistic. Group members who have participated in HFH have all stated that prior to working the construction supervisor conducts a safety brief that can last up to 30 minutes. Our concern is not with the length of the safety brief, but with the quality of it.

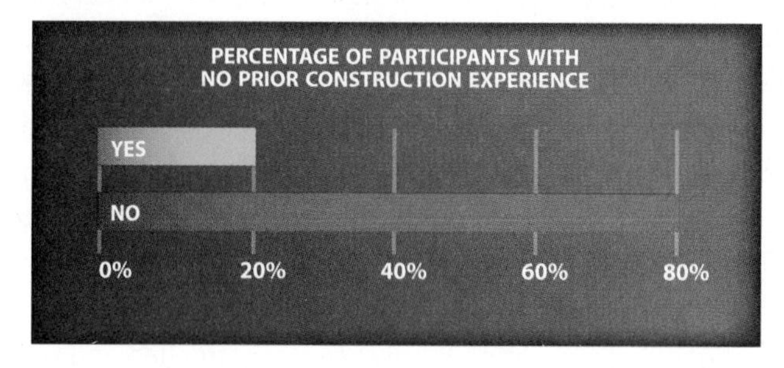

Figure 1. Survey feedback: volunteer construction experience. *Source: "Habitat for Humanity Survey." Survey. Surveymonkey.com. 14 Nov. 2012.*

In order to be certain everyone is getting an effective insight to the safety practices in the construction workplace, safety training should be consistent and uniform. Our proposal to improve safety and to enhance the comfort and confidence of those working around volunteers is to have all volunteers participate in a mandatory online safety course. This safety course would provide an excellent source of pertinent general safety information and include a test in which participants would have to pass in order to volunteer. The good news is that this online safety course already exists. It is part of an insurance program administered by Lockton Risk Services, Inc. and works with the Habitat Affiliate Insurance Program. The site not only includes the general safety course, which we are proposing to be mandatory, but also over 75 other courses for specific safety procedures and tool training. This brings us to our next area of proposed improvement.

Tool Training

HFH does require that a volunteer is comfortable in using a tool before operating it. According to Kelsey, the construction team makes an effort to identify those who have experience with a specific tool and train those volunteers who want to learn to use a certain tool. The HFH construction teams are very dedicated people and are always concerned with the safety and well being of all the volunteers. They also take pride in helping the new home recipients gain knowledge on proper tool use in an effort to contribute to the soon to be home owner's capability of maintaining the home themselves. While a lot of effort is being put into this, it is apparent that it takes away from the construction teams efforts to focus on the construction process themselves.

In an effort to increase the efficiency of the construction process and to prevent delays due to training, our group proposes that volunteers take online tool courses that focus on the proper and safe use of each tool. HFH will then certify each volunteer and allow that certification to be entered in that volunteers records so that HFH can use that information to invite volunteers to future builds that may require the use of that tool and skills those tools help encompass. This will create a databank that will help the HFH construction team utilize volunteers more efficiently. We also believe that this, like the website discussed in the "Volunteer and Scheduling Reliability" section, will help develop the relationship between HFH and its volunteers. Volunteers will be more willing to learn and want to contribute more by trying to become better qualified so that they may be of better service to HFH.

Decrease in Housing Values around HFH Housing

One of the effects of building Habitat for Humanity (HFH) homes in a community is the risk of a decrease in property values. It is important to analyze the impact of HFH homes on the property value of other homes because property value is significant in determining what the market price of one's real estate. Furthermore, property value also determines how much one must pay in local property taxes, which can affect one's overall monthly payments. According to the *Commonwealth of Kentucky's Department of Revenue* office, "property tax in Kentucky is levied on the fair cash value of all real and tangible property unless a specific exemption exists in the Kentucky constitution ..." The fair cash value of one's property is the amount for which a property can be sold in the due course of business and trade, not under duress, between a willing buyer and a willing seller (*USLegal.com*). This therefore makes property value an important issue to consider when building or buying a new home, especially low-income homes as is the case of HFH.

Amy Handlin, a contributor to *ehow.com,* states in her article "What Makes a Property Value Decrease" that of one's property and that of one's neighborhood's property causes one's property value to decrease as a result of it being poorly maintained. Other factors such as the health of the local economy, the land and location factors, the trends sweeping society (such as the Green movement) and the city, county, and state government also play a role in determining the value of one's property. However, in regards to the Lexington HFH homes and around the nation, many people have claimed that building a low income, charitable home next to their regularly paid mortgaged home has caused their homes to devaluate. Because HFH builds low-income homes for families in need whose incomes are below 60% of area median income for Fayette County, they themselves rely on charitable donations and volunteers to build these homes. This claim is thus based on the fact that the low-income homes built by HFH causes other home values in the same neighborhood to denigrate.

Non HFH residents, especially those in exclusive areas, fear that an increase of criminal activities, traffic, and an overall "cheap look" of the neighborhood can be triggered by building low income homes in their neighborhood, and thereby, lower the value of their properties. This fear has caused many residents to resist to the idea of building low income homes in their neighborhood and have thus coined these low income projects, including HFH homes, a term known as NIMBY "Not in My Back Yard" developments. According to *http://www.dictionary.com,* a NIMBY development is something that, "though needed by the larger community, is considered unsightly, dangerous, or likely to lead to decreased property values."

Homeowners consider these NIMBY projects as the main reason that causes property values to devaluate because of the fact that appraisers use other homes in the neighborhood as "comparables" in order to determine the value of a property. According to *The appraisals foundation,* an ad hoc Committee on Uniform Standards of Professional Appraisal Practice, there are two primary appraisal methods for residential property: a sales comparison approach and a cost approach. In the sales comparison approach, an appraiser compares the property with three or four similar homes that have sold in the area. In the cost approach, new properties are analyzed based on reproduction costs. The appraiser estimates the cost to rebuild the home if the property were destroyed and then looks at the land value and depreciation to determine the property's value.

HFH Interviews

In order to find evidence and further clarity about these claims, we conducted a number of interviews with people affiliated to HFH Lexington. The first interview was with a HFH home recipient, Princia Itoula, who revealed that she was aware of the claims that neighbors had about HFH homes. Itoula claimed that her neighbor was very territorial but did not know of any specific details concerning property value. The second interview on this subject was with the HFH Family Services Director, April Smith, who claimed that she was aware of the fact that people accused HFH homes causing their home values to decrease. Although Smith did not deny those claims, she indicated that this was not an issue for the Lexington HFH chapter. Lastly, we conducted an interview with George Joudeh, a neighbor of a HFH home, who claims and can prove that an HFH home in his neighborhood did indeed decrease the value of their homes.

INTERVIEW WITH HFH NON-RESIDENT

In a brief interview with George Joudeh, a non HFH resident who claims that HFH built a home in their community, which instigated the decline in property value in their neighborhood and as a result, some homes lost between $10,000 to $14,000 in property value. Mr. Joudeh explained that a neighbor's home burned to the ground and they decided to donate the property to HFH, who consequently built the low-income home that caused the devaluation of their homes. He went on to say that although HFH is a good program, the properties in their area were already cheap, so "it's not nice that home values went even lower than what it was already" (Joudeh).

INTERVIEW WITH HFH DIRECTOR

When we interviewed April Smith, the HFH Family Services Director, one of the questions we asked was whether HFH homes reduced the value of homes in the surrounding area. According to her, the Lexington HFH chapter does not have that problem but she was aware that there were such claims in other states. Smith explained that HFH Lexington targets distressed areas as potential and eventual areas in which they build their homes and, therefore, a declination in property values does not occur as a result of HFH homes but rather because of other reasons such as property and neighborhood maintenance. This fact is supported by Amy Handlin's article "What Makes a Property Value Decrease" mentioned earlier. In that article, the author states that property and neighborhood maintenance are the main reasons why properties lose value because housing values revolve more around the condition of the actual home being sold or bought. Smith also explained that the maximum monthly mortgage payment is about $400 a month for each HFH home and clarified that this is a strategic move made by HFH in order to help their home recipients be able to afford the maintenance and upkeep of their properties. But according to Joudeh, these low home values and monthly payments HFH requires are exactly the reasons why his homes lost its value because appraisers have compared neighborhood homes to HFH homes.

In regards to the maintenance and upkeep of their homes, April Smith explained that all of their applicants go through extensive training about how to become a homeowner and how to maintain a home so that once they receive their homes, they would be more than qualified to manage it. Joudeh on the other hand, claims that one cannot be trained to change ones habits in a few hours and therefore, even though HFH trains their home owners, the demands of maintaining and managing a home is a quality that one should have during one's upbringing and, he therefore claims that HFH residents don't maintain their properties as they should since it is their first time to have such an opportunity. He therefore claims that overall property values decline as a result of HFH projects.

INTERVIEW WITH HFH RESIDENT

In relation to the claim that HFH homes lowers the home values of other homes, an interview with a HFH home recipient was conducted to get an idea about how much training they had in regards to home maintenance. Princia Itoula was interviewed and informed us that she had received approximately 500 hours of home maintenance training required by HFH. Furthermore, she said that the 500 hours also serve as a down payment towards the home since HFH does not require one. Given this information, we as a group feel that this is not an effective way to train people because the focus of the applicant would be to get their application approved which can therefore result in them going through the motion of things during training. Additionally, we feel that the fact that most of these applicants have never owned homes and come from disadvantaged backgrounds makes home maintenance training a very important part of home ownership in order to preserve the value of their property.

In the interview with Itoula, we learned that HFH residents who no longer want to keep the HFH home or continue with their application when their home is being built can sell the home back to HFH and, HFH resells it to other applicants. This was the case for this HFH resident who bought the home when it was already almost complete and as a result HFH sold her the Home for $30,000 less than what the previous applicant had paid for, and this home had never been inhabited. Given this information, we concluded that HFH homes have a high potential of lowering the home value of other homes not only because of the volatility of their home prices, but also because of the low values and low quality of the homes.

Investigating the Claims

To investigate Joudehs' claims we conducted our own research about where HFH built their homes. So far this year, HFH has issued 15 homes, each of which had a different sponsor, including the University of Kentucky's HFH campus chapter, interfraternity council, and the Pan-hellenic council, which sponsored the property on 450 Ash Street. Other sponsors for this year's HFH homes were organization such as Lexmark, Webasto, Fifth Third Bank, and Calvary Baptist Church, just to name a few.

According to their website, HFH claims, "each home built with Lexington Habitat helps transform not only one family's life but changes the street, the neighborhood and our community as a whole. Improved streetscapes, safer structures, increased property tax base and an infusion of resources into the local economy affect the quality of life for all of us." Several neighboring homeowners, such as George Joudeh, disagree with these claims because, not only do the low-income homes devaluate other properties, but also because of the type of people these types of projects tend to attract. We took one sample of the 15 homes HFH dedicated this year and compared it to four other properties in the area as illustrated below in Figure 2.

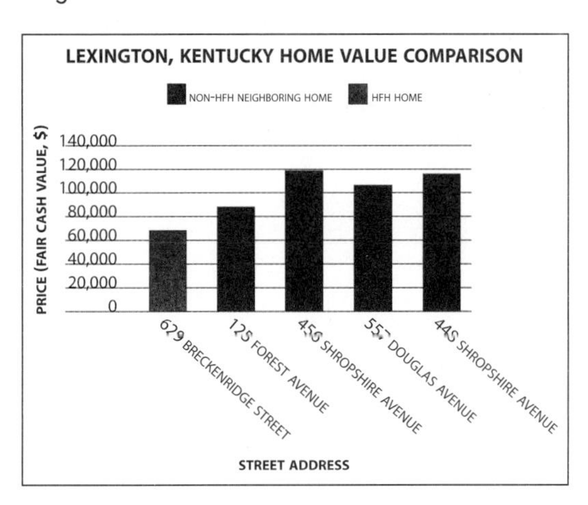

Figure 2. Comparison of HFH home to neighborhood homes. *Source: Fayette County Auditor.*

In the illustration on the previous page, the HFH property on 629 Breckenridge Street is compared to four other properties in the same area and, as one can see, the HFH property valued at $68,600 is well below the property value of the other houses, which range between $88,000 and $118,900. Based on this information, we can therefore conclude that Joudehs' claims are valid and that HFH properties can cause other houses to devaluate.

Suggested Solution

The best possible solution we as a group have come up with is for HFH to apply for an Inclusionary Zoning Affordable Housing Program with the city ("Inclusionary"). According to the *Department of Housing and Community Development*, IZAHP refers to "a land use technique for developing diverse mixed-income communities by requiring each new residential development make a percentage of the new units affordable to targeted incomes" ("Inclusionary"). The incentive in this program is for those community developments to receive a Density Bonus, which according to *the Center for Land Use Education* is a program whereby, developers are permitted to increase the square footage or number of units allowed on a property if they agree to restrict the rents or sales prices of a certain number of the units for low income families. As discovered by the Center for Land Use Education, the additional cash flow from Density Bonus units offsets the reduced revenue from the affordable units (Miskowiak and Stoll). This therefore supports our claim that low-income homes reduce the value of properties in surrounding areas.

Since the goal of the IZAHP is to create mixed income neighborhoods; produce affordable housing for a diverse labor force; seek equitable growth of new residents; and increase homeownership opportunities for low and moderate income levels, we believe that both HFH and non HFH residents can benefits from this goodwill act because it will reduce the "nimby" attitudes and non HFH home owners can have the opportunity to be a part in the decision making process. Table 1 below gives examples of the different types of cost offsets homeowners can benefit from IZAHP.

Table 1. Types of cost offsets found in IZAHP. *Source: "Inclusionary Zoning Affordable Housing Program." Dhcd.dc.gov.*

Type of Cost-Offsets	What It Does and Why It Helps Developers	Example
Density Bonus	Allow developers to build at a greater density than residential zones typically permit. This allows developers to build additional market-rate units without having to acquire more land.	Most jurisdictions offer density bonuses. Typically they are equivalent to the required set-aside percentage.
Unit Size Reduction	Allow developers to build smaller or differently configured inclusionary units, relative to market rate units, reducing construction and land costs.	Many programs allow unit size reduction while establishing minimum sizes.
Relaxed Parking Requirements	Allows parking space efficiency in higher density developments with underground or structured parking: reducing the number or size of spaces, or allowing tandem parking.	Denver, Colorado, waives 10 required parking spaces for each additional affordable unit.
Design Flexibility	Grants flexibility in design guidelines—such as reduced setbacks from the street or property line, or waived minimum lot size requirement—utilizing land more efficiently.	Boston, Massachusetts, grants inclusionary housing projects greater floor-to-area ratio allowances.
Fee Waivers or Reductions	Reduces costs by waiving the impact and/or permit fees that support infrastructure development and municipal services.	Longmont, California, waives up to 14 fees if more affordable units (or units at deeper levels of affordability) are provided.

Table 1. (Continued) Types of cost offsets found in IZAHP. *Source: "Inclusionary Zoning Affordable Housing Program." Dhcd.dc.gov.*

Type of Cost-Offsets	What It Does and Why It Helps Developers	Example
Fee Deferrals	Allows delayed payment of impact and/or permit fees.	San Diego, California, allows deferral of Development Impact Fees and Facility Benefit Assessments.
Fast Track Permitting	Streamlines the permitting process for development projects, reducing developers' carrying costs (e.g., interest payments on predevelopment loans and other land and property taxes).	Sacramento, California, expedites the permitting of inclusionary zoning projects to 90 days from the usual time frame of 9–12 months. Savings average $250,000 per project.

As illustrated above, IZAHP can benefit HFH because it not only helps them solve the property devaluation issues, but also other issues such as permitting, fee deferrals and even fee waivers, which were all issues that April Smith had mentioned. For this topic we conclude that IZAHP can help HFH's mission for helping "Families in need achieve homeownership by purchasing simple, decent affordable homes they help build with community partners" (*Lexington*).

Building versus Repairing

In 2008, the housing market in America hit an all time low. An unprecedented boom in the market in the early 2000s hit a wall in 2006. Average home prices began dropping, financing became more difficult, refinancing became near impossible, and mortgage payments rose creating an economic collapse. Homes were depreciating instead of appreciating and becoming worth much less than buyers had anticipated. Homeowners were forced to decide between attempting to sell at a loss, renegotiating with lenders, or failing to make payments, which leads to foreclosures and defaults. RealtyTrac, a program that lists national housing information, states that over 3 million homes were repossessed between 2007 and 2010. Today over 11% of all housing units are vacant (Olick). The S&P/Case-Shiller report below demonstrates the housing market collapse. See Figure 3 for a visual representation of the steep drop in the average price of homes.

MEDIAN HOME PRICES AFTER CRASH

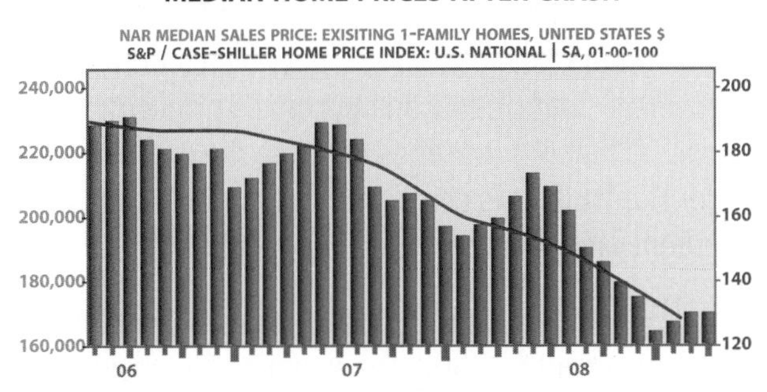

Figure 3. Median home prices after crash. *Source: National Association of Realtors and S&P Dow Jones Indices.*

As you can see, average home prices were at a high in 2006 because of the housing market boom. As time progresses, prices start to fall. In 2007 the prices start a steep decline and bottom out around 2009, when the housing market was at its worst.

Housing Market Today

The housing market today is on its way to recovery, but still not in great shape. According to the Economist, home prices are still on average 19% below fair value. This 19% represents the fact that homes are not holding their values and are selling for prices much less than what they should actually be worth. Rochelle Fitzgerald, a sales associate with Coldwell Banker Residential Brokerage, said "I would caution buyers to be careful buying a new home because builders are competing against foreclosures and it could be a long time before a new home will increase in value" (Learner). Since home prices are so far below value, would-be-sellers are having trouble marketing their homes to potential buyers because buyers are obviously attracted to lower prices. To be able to compete, sellers are forced to lower the price of their homes, thus creating the lull in the housing market we see today. It's a simple supply and demand example. Because there is such a large supply of homes and a limited demand, prices are being driven down. Until there is a significant increase in demand it will be difficult for values and prices of homes to rise.

Repair Program

Habitat for Humanity's business model is to build homes for those that cannot finance on their own. But with the housing market experiencing a sharp decline, are more homes really needed? To be able to purchase a house through Habitat for Humanity, applicants must go through personal finance training and once the applicant acquires the home HFH becomes the mortgage holder. This would take care of any possible credit/mortgage issues in the future from HFH homeowners. So if credit won't become a problem, it seems the problem with building more homes is that it is just does not have the demand that there once was. A proposal for Habitat for Humanity is to shift some of their focus from building completely new homes to repairing/renovating currently existing homes.

According to Habitat for Humanity's website, the costs to build a home differ between regions but usually average at about $70,000. The average cost to repair or remodel a house is between $40 and $50 per square foot. Homes built by Habitat for Humanity are no larger than 1,050 square feet. So if HFH were to repair or remodel homes this size, costs would be only be $42,000–$52,500 per home which is $17,500–$28,000 cheaper than building from scratch. However, a large majority of the cost to repair or remodel is labor. Habitat for Humanity gets it labor from applicants and volunteers, which is essentially free labor. So if only half of the $40–$50 per square foot cost is labor, then the savings are even higher for repairing/remodeling.

In the survey we created and presented to our peers, a question about a repair program was included. The final question of the survey was "Would you be interested in volunteering for a Habitat program that repaired houses instead of building new houses?" The results are shown in Figure 4 below.

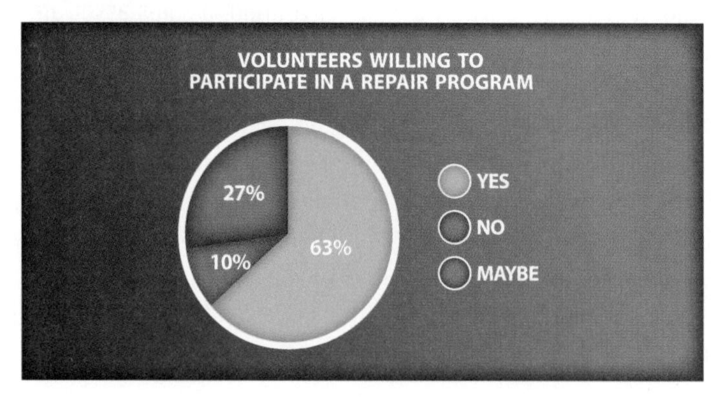

Figure 4. Survey Feedback: Volunteer Willingness. *Source: "Habitat for Humanity Survey." Survey. Surveymonkey.com. 14 Nov. 2012.*

Of the total 30 responses to the survey, 63% said they would be interested volunteering for a repair program. Roughly 27% said maybe and only 10% said no. If 63% of HFH's current volunteers would be willing to shift their work from building new homes to a repair program, this would be more than enough labor needed to design and execute this new program.

A Brush with Kindness

Upon further research, Habitat for Humanity does indeed have a program dealing with home repairs, but it is very new and in the very early stages of creation. The program is called "A Brush With Kindness" ("Habitat for Humanity"). The work from ABWK would focus on essential external repairs and improvements to existing homes of qualified, very low-income homeowners ("Habitat for Humanity"). The program is aimed to aid seniors, people with disabilities, and others that are financially unable to make essential repairs to their homes. While this is a great step in the right direction, it can be improved.

While A Brush With Kindness focuses on external repairs that are urgently needed, a new program could be created that focuses on internal repairs and remodeling. As discussed before, the demand for new homes is small, but the need for repairs or remodeling is not. These internal repairs would be aimed to fix home deficiencies such as faulty plumbing, mold, holes in walls, flooring flaws, etc. These repairs would upgrade current housing for those in need without building complete new homes. This program would follow other Habitat for Humanity programs in that it would utilize volunteer labor and donated funds for materials. Repairing and remodeling homes would not only serve to help families, but it could also slightly aid the housing market's recovery. Since most of HFH's work is done in "distressed communities," remodeling these homes could help raise their values.

Conclusion

Habitat for Humanity has undoubtedly changed the lives of thousands over the years. With the low cost housing solutions that they provide, people who were previously living in very poor conditions can reside happily in their newly constructed home while not having to fear a large mortgage payment that they can not make each month. This program exemplifies the ideal charitable organization just on the volume of people assisted alone. With all of this being said, if the problems above could be remedied, the effectiveness of Habitat for Humanity would reach a new level of service and effectiveness. After the extensive research that we have just presented, we recommend the following changes:

- Keep a certification database of trained volunteers.
- Create a volunteer scheduling website to access more volunteers.
- Utilize the Inclusionary Zoning Affordable Housing Program.
- Further develop a program that focuses on internal repairs and remodeling for existing homes instead of new homes.

While the presented changes are not major, improvements in the organization will be seen nationwide. Overall, most of the suggestions that we are making have to do with shifts in modern-day economics as well as advances in technology. By doing things such as implementing a new website or keeping a database of volunteers, you are embracing new technologies to take Habitat for Humanity to the next level. Similar to this, by recognizing that the housing economy is different from what it was a decade ago, our proposed housing repair plan can be used to customize Habitat for Humanity to the current market. Every organization has to strive to keep up-to-date with how society as a whole is constantly changing. We hope this report has aided in this adaptation and we look forward to the benefits that you will see as a result.

Works Cited

"The Appraisal Foundation." AppraisalFoundation.org. N.p., 2010. Web. 05 Dec. 2012.

Collins, Michael. "Dr. Boyce Real Estate." *: What Does "NIMBY" Mean?* Dr. Boyce Real Estate, 23 Apr. 2009. Web. 05 Dec. 2012.

"Planning Implementation tools Density Bonus." *University of Wisconsins Stevens Point.* Center for Land Use Education, Nov. 2005. Web. 05 Dec. 2012.

"Fair Cash Value Law & Legal Definition." *Fair Cash Value Law & Legal Definition.* N.p., 2012. Web. 05 Dec. 2012.

Fayette County Auditor. *http://www.fayette-pva.com.* Web. 13 Jun. 2013.

Giauque, Kelsey. Construction Coordinator. Lexington Habitat for Humanity. Personal Interview. 3 Dec. 2012.

"How To Use It." *Inclusionary Zoning-PolicyLink*. Policy Link, 2003. Web. 05 Dec. 2012.

"Inclusionary Zoning Affordable Housing Program." *Dhcd.dc.gov.* Department of Housing and Community Development, n.d. Web. 05 Dec. 2012.

Itoula, Princia. Personal interview. 13 Nov. 2012.

"Habitat for Humanity to Offer Essential Home Repairs to Qualified, Very-low-income Homeowners." *Thedailyreview.com*. The Daily Review, 2 June 2010. Web. 26 Nov. 2012.

Joudeh, George. Personal interview. 23 Nov. 2012.

"Lexington Habitat for Humanity." *Lexington Habitat for Humanity.* N.p., 2012. Web. 05 Dec. 2012.

Lerner, Michele. "Should You Buy a Newly Built Home?" *MSN Real Estate*. MSN, Aug. 2011. Web. 26 Nov. 2012.

Lockton Risk Services. Habitat for Humanity. Volunteer Safety Training. Nov. 2012. Web. 18 Nov. 2012.

"Median Sales Price of Existing Single-Family Homes for Metropolitan Areas." National Association of Realtors. 2013. Web. 13 Jun. 2013. *http://www.ralsc.org/marketing_pdf/Median%20Sales%20Price%20of%20Existing%20Single%20Family%20Homes.pdf*

Olick, Diane. "Nearly 11% of US Houses Empty." *Realty Check*. CNBC, 31 Jan. 2012. Web. 26 Nov. 2012.

"Neighborhood Stabilization Program Grants." *HUD.GOV.* U.S. Department of Housing and Urban Development, n.d. Web. 26 Nov. 2012.

Parrington, Tami. "The Effects of Low Income Housing on Surrounding Property Values." *EHow.* Demand Media, 29 July 2010. Web. 05 Dec. 2012.

"Property Tax." *Kentucky: Department of Revenue*. N.p., 2012. Web. 05 Dec. 2012.

Rogers, Jaimee. "What Brings Down a Property's Value?" *EHow.* Demand Media, 30 Apr. 2009. Web. 05 Dec. 2012.

Siddons, Sarah. "How Home Appraisals Work." *Home.HowStuffWorks.com*. Howstuffworks, 05 Aug. 2008. Web. 05 Dec. 2012.

Smith, April. Personal interview. 16 Nov. 2012.

"S&P/Case-Shiller 10-City Composite Home Price Index." S&P Dow Jones Indices. McGraw Hill Financial. *www.spindices.com/indices/real-estate/sp-case-shiller-10-city-composite-home-price-index*. Jun. 2013.

Washington, R.A. "The Rebound Is Now." *The Economist*. The Economist Newspaper, 5 Apr. 2012. Web. 26 Nov. 2012.

"What Is the Average Home Remodeling Cost per Square Foot?" *Calfinder.com*. Calfinder, n.d. Web. 26 Nov. 2012.

Appendix

Survey Questions

1. How many times have you participated in Habitat for Humanity?
2. What organization did you volunteer with?
3. What tasks were you asked to perform?
4. How many workers were employed by Habitat?
5. How much training were you given on site?
6. Did you ever perform tasks you did not feel qualified for?
7. Did you witness any work being done that you thought was low quality?
8. How would you describe the neighborhood that the house was being built in?
9. Would you be interested in serving in a Habitat program that fixed houses for people instead of just building them?
10. Do you believe that Habitat should be responsible for repairs to their homes?

Interview: Family Services Director, Habitat for Humanity

Name: April Smith, the Family services director at HFH Lexington

Date: Friday, November 16, 2012

Subject: Overview of HFH

Questions:

1. So they guarantee the house for a year, but after that year is up, do they offer any house servicing, repairs, remodeling, etc.?

   One year warranty.

   No house servicing because each recipient goes through an intense home keeping training.

   The goal is self-sufficiency and not to hold their hand forever.

2. For the repair programs that they do offer, is it for overall remodeling (full re-haul of the house), general repairs, or what is it they offer?

   It's a new program.

   For remodeling check out the Neighborhood Stabilization Program. It's federally funded through LFUCG.

   LFUCG got properties and passed on to HFH.

3. Do they offer any kind of discounts on home repairs to the houses they have built in the past? For example, someone buys a HFH house and needs repairs four years down the road, does HFH offer a repair program or discount on repairs for that home?

   Cost of homes is based on an appraiser's rate.

   Self-sufficiency is their goal.

   After an applicant acquires a home, HFH becomes mortgage holder at that point.

4. Do Habitat for Humanity homes bring down the value of homes in the surrounding area?

   According to April Smith, no. because they build the homes in distressed communities.

   $400 is the avg. monthly payment. It is made affordable for home so that home owners can afford to do any repairs.

5. Do they employ any construction personnel like an experienced foreman? Ever?

   No, only contractors such as electricians and plumbers.

   Otherwise they do all the work themselves. It just takes a lot of planning and time to get permits, surveying, house plans.

   They have a "constructions department" that does all that as well as order materials, book contractors, electricians, plumbers, etc. The construction department has to be on site at all times.

   As a follow up question, I asked if the "construction department" is licensed or have any certification. Smith said no. They are first aid and CPR certified though as well as OSHA trained.

6. What, if any, kind of construction or safety training do they give?

   There is a safety talk every morning.

   They are working towards having training videos, which Habitat international provides—material, and training videos.

   She said HFH has a good safety record.

7. When it comes to the construction process, how exactly does the foundation, plumbing, and electric work get installed? Are they contracted out?

   Some is done themselves.

8. What are some (common) problems that occur during the construction process?

   Delays, needing million permits. For example, the framing has to be done at right time, which delays everything else. This creates a snowball effect. Weather. City requirements. LFUCG.

9. How often do homeowners run into problems due to poor construction and what are these problems, if any?

   Occasionally there are minor plumbing problems. The warranty is with the plumber.

   This is referred to as "callbacks."

10. Who does the final inspection? Is this legally official?

    Yes, it is legally official. LFUCG does the inspection; a certificate of occupancy is then given before a family moves in.

    The remodeling program is a fairly new program (started this year only), see the Neighborhood stabilization program.

11. Safety statistics?

    No stats.

Interview with HFH Neighbor

Date: Friday, November 23, 2012

Name: George Joudeh, Neighbor to HFH home

Subject: Do you think HFH homes decrease the property value of other homes?

Goal: To get evidence (factual or opinionated) from a HFH neighbor that HFH homes does bring down home values.

After a brief introduction and description of the project, the following questions were asked and answered by George Joudeh.

Questions

1. What price range are the homes in your neighborhood?
2. How would you describe your neighborhood?
3. What do you know about HFH?

    They are a not for profit organization that build low income homes in various neighborhoods for low income families.

4. How do you feel about them building a home in your neighborhood?

   "I think, HFH is a good program, but properties in this neighborhood are already priced low, and for HFH to come and build homes here is not good because we experienced a 10 percent decrease in our property."

5. How did you find out that property prices had decreased?

   Neighbor wanted to refinance their home for remodeling. Their property was re-appraised and they were informed that their home had lost its value and they pointed to the HFH home as the reason.

6. HFH says that they build homes in distressed areas is this true? Is your neighborhood in distress?

   That is what HFH claims, but in reality they can build a home anywhere they want because not only are there no laws to stop them, but they receive properties through different forms of donations, which was the case for this home.

   The previous home had burned to the ground and as a result, the previous owner donated the land to HFH who then built a new home.

   This is not a distressed area. I'm originally from Michigan and we can easily say that this is a decent neighborhoods; $120,000 homes are not built in distressed areas.

7. What would you advise HFH to do?

   Although no one can stop them from building a home, HFH should reevaluate their definition of distressed and at least hold neighborhood meetings before building homes .

Interview with HFH Resident

On Tuesday November, 13th, we conducted an interview with Princia Itoula, a Habitat for Humanity resident who lives on Georgetown road, Lexington, KY.

Since we know the HFH resident, no formal interview questions were formulated , rather we just asked her to talk to me about her experience with HFH.

According to Miss Itoula:

- HFH houses are built by volunteers only and then when the construction is finished, a contractor is called to inspect the home.
- The majority of volunteers are the current HFH applicants who are obligated to fulfill a minimum of 500 hours per adult in the family. Therefore, before your home is built, you have to fulfill the 500 required hours which is seen as your "investment," (all current applicants build each other's homes).
- These 500 hours also serve as a "down payment" for the home you will receive since actual down payments or interests are not charged.
- Each applicant has a "sponsor," who could be a church organization, a university, or a bank, which agrees to fund the home (give a home loan).
- Miss Itoula's sponsor is a Baptist church.
- The number of bedrooms in each home is built according to the number of household members, but generally, with a maximum of 4 bedrooms.
- The maximum annual gross income to qualify for an HFH home is $27,000.
- Each applicant is required to have a job.
- A credit check is conducted to see if the applicant is (overwhelmed) in debt.
- If so, then HFH draws a plan to help the applicant pay off all debts before starting the HFH home process.
- HFH have orientation classes on homeownership.
- All homes are guaranteed for one year only.
- HFH also buys homes and repair/remodel's them.
- The "sponsors" buy the homes and offers them to HFH applicants.
- HFH applicants are also offered homes from current HFH residents who no longer wish to stay in those homes.
- The whole HFH home process takes between 8 to 18 months.
- When current HFH homes are resold, they sell for a very low value.
- HFH monthly home payments are determined by the income at the time of application, and are not adjusted afterwards, regardless of whether the residents earn more or less.
- In regards to claims of the property value decrease of other homes as a result of HFH homes, Itoula is aware of claims but does not have any further information. She claims that some neighbors are very cold and territorial towards her and her family.

Organizing Your Thoughts (organization patterns)

Although there's not one "correct" way of writing an essay, there are some organizational strategies writers can use to help them organize their ideas. The following are some very popular organizational patterns that are useful for organizing essays, speeches, or any kind of project that needs to have a coherent structure:

- **Chronological:** This pattern of organization is useful for describing events that happened along a certain timeline. For example, if you're talking about the history of desegregation on the University of Kentucky's campus, you might want to organize your essay by year. You might also think about chronological patterns for describing any kind of events that happen in a series.
- **Spatial:** This pattern of organization is most useful for describing the locations and comparable distances of various artifacts. For example, people who argue that Rupp Arena needs to be improved often begin by describing the current structure of Rupp—from parking lot to entry points to bathrooms to the court itself.
- **Comparison:** This pattern of organization is useful when you are comparing two or more items. You may identify several points of overlap or difference that should be illustrated, and you can organize your paper around those points. For example, if you want to argue that a vegetarian diet is healthier than a non-vegetarian diet, you might choose to highlight several points of comparison:
    - Heart disease: vegetarians versus non-vegetarians
    - Average time lost to sick days: vegetarians versus non-vegetarians
    - Cost of groceries: vegetarians versus non-vegetarians
    - Obesity rates: vegetarians versus non-vegetarians
- **Cause/effect:** This pattern of organization is useful for showing a causal link between an event (or series of events) and the effects that came as a result. You do not necessarily need to begin with the cause, although beginning with an explanation of the cause would certainly make sense. You might want to start with a strong description of the effects and then proceed to reveal the causes. For example, a writer might begin by describing the terrible car wreck she was involved in: the mangled car, her injuries, and the financial aftermath. She might then reveal some of the things that helped contribute to this crash: the driver had been texting a friend, driving too quickly on a dark road, and not giving herself enough time to slow down before an intersection.

- **Problem/solution:** This pattern of organization is very common for writers who are arguing for certain actions to be taken. In this pattern, you would first explain the problem in detail for your readers. Show what has led to the problem, what negative effects it has, and so forth. Next, you present your solution or answer to this problem. It may be that you want your audience to change its mind about something, or you want to motivate them to take action.
- **Topical order:** This pattern of organization is useful for writers who want to organize their writing around particular topics that are relevant to their larger point. The topical patterns of organization may be most useful for informative writing that describes or tells readers about the various aspects of a bigger subject. For example, if you are informing readers about types of higher education institutions, you might organize around the following topics:
    - Public universities
    - Private, liberal arts universities
    - Community colleges
    - Two-year public colleges

Writing: Some Techniques

After you've brainstormed your ideas and planned them out, it's time to start writing. There is really no one right way that writers do their work. Some people love to sit in front of a computer in total silence, while others love to work with pen and paper in a coffee shop. Still others might have different ways to get their best work accomplished. Try some of the following methods and compare how you feel. Do you find yourself working with more focus in one setting? Do you find that certain sounds distract you more than others?

Writing with Headphones

Headphones might seem like a strange writing tool, but many people find that music can help generate just the right mood for writing and thinking. You might need to experiment with the right type of music for getting work done. Of course, you might be one of those people who finds writing with music altogether too distracting. Try several different settings and note the results.

Writing at Different Parts of the Day

Professional writers often can tell you what times of day they prefer to write. I am a "morning writer," for example, because I love waking up early and getting my writing done. Other writers are passionate about "night writing" because this is when their creative juices start flowing. Depending on work schedules, it's not always possible to write *exactly* when we feel ready. But you might try experimenting with writing at different parts of the day. How do you feel when writing in early or late afternoon? Do you like writing late at night? After you sit down for a session, make a mental note of how it felt to write at this time.

Write for Certain Periods of Time

It's almost never a good idea to sit down and write an essay (or anything else) in a single session. Some students wait until the night before a paper is due in order to write it. The adrenaline and panic usually help writers to produce the required number of pages, but the quality is not usually the best. As you move from high school to college-level writing, the hurried and sloppy quality starts affecting your grades.

Instead of sitting for hours at a time in front of the computer, you can save yourself a lot of stress by breaking up the task into smaller pieces. Some writers and researchers have even suggested that giving our brains set boundaries and deadlines can help to increase focus. The "Pomodoro Technique" has become quite popular among many writers for just this reason. Here's how it works:

1. **Get a timer:** You can use an online timer, your phone's timer, or an actual timer.
2. **Set your timer for twenty-five minutes:** Write non-stop for twenty-five minutes. Don't get up to walk around. Don't stop to check your email. Just write for twenty-five straight minutes.
3. **When the timer goes off, take a full five-minute break:** Don't do any work. Now you can check your email, get some coffee, or just close your eyes. Take the full break.
4. **After your five-minute break is done, set the timer for another twenty-five minutes and write again:** Then take another five-minute break.
5. **Repeat the cycle for as long as you want to work.**

Revising

After writers complete a rough draft, the work of revision begins. Revising is not the same as editing or proofreading. In fact, revising is much more creative, critical, and exciting. During the revision process, many writers literally re-visualize their project. They get feedback from others, read their own work, and try to imagine ways of expanding on places that need more depth. Some of the following revision questions are worth asking yourself when starting your revision:

- Does this draft actually fulfill the assignment? Have you strayed at all from what you were asked to do? Look back over the assignment prompt and try to determine whether or not you've managed to fulfill all parts of the assignment.
- What is your main point or central argument? Does the entire text actually support, explain, and make the case? Do you ever drift into unrelated tangents that don't exactly relate to your main point?
- Are there parts that seem confusing or unclear? Ask your readers to tell you if they get confused or lost at any points. If so, how can you revise those sections so that they address your readers' confusion?
- Are you able to clearly follow your own organizing structure of the whole paper? Try to make a "reverse outline" of your paper.
- Do you have an introduction that grabs your readers' attention? Does it set the tone for the rest of your work?

Editing: Lanham's Paramedic Method

In his excellent book *Revising Prose,* Richard Lanham introduced what he calls the "paramedic method" for editing your own work.[2] This method is a great way to help find those awkward sounding sentences that might confuse (or annoy) readers. Lanham's method is designed to give your writing more punch, more power, more persuasion. It's designed to make your prose more exciting to read. By following the paramedic method, you will eliminate boring, sluggish sentences and unnecessary wordiness. Here are the steps:

1. **Circle the prepositions.** Prepositional phrases often create wordiness and some s-l-o-w reading. (Examples are: around, at, by, for, from, in, of, on, over, outside, through, since, under, up, with, etc.)
2. **Draw a box around the "to be" verbs.** This is a good way to catch passive voice construction that detracts from exciting, active prose. (For example: am, is, are, was, were, etc.)

3. **Ask yourself, "Where's the action?" and "Who is kicking whom?"** Figure out who is actually performing the action in your sentences. Who or what is the main subject of each sentence? Revise each sentence to clearly show the actor in charge.
4. **Change the action into a simple verb.** Now that you know who is doing the action (who is doing the kicking!), revise each sentence to give that actor a simple action verb.
5. **Eliminate the slow wind-ups at the beginning of sentences.** Wind-ups are phrases like, "I believe that ..." and "In my opinion ..." These kinds of beginnings aren't necessary and they slow down the reader. Instead, jump right into the sentence.

Revising on Screen versus Revising on Paper

Some writers find it helpful to print out a draft on paper in order to revise and edit. Revising is sometimes easier to do with pen and paper, rather than online. Try it both ways to see which one you prefer.

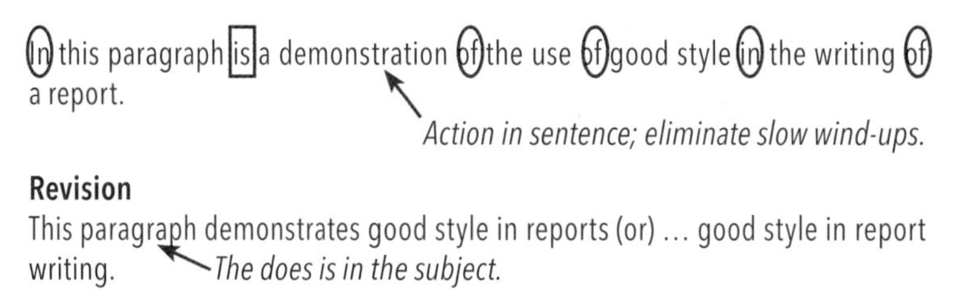

Figure 5. Example using the Lanham's method.

Example: Currently there ***is*** a project that ***is*** being done **to** help students **in** after school programs **to** further their learning and have educational resources made available **to** them.

- Who is doing the action? Who is kicking whom? *"A project"*
- What is the action? *Helping, furthering, and making*

Revised sentence: Currently, a project is helping students in after school programs by furthering their learning and making resources available to them.

Example: The point I wish **to** make ***is*** that the employees working **at** this company ***are*** **in** need **of** a much better manager **of** their money.

- Who is kicking whom? *Employees*
- What is the action (the kicking)? *Needing*
- Slow wind-up? *The point I wish to make is that …*

Revised sentence: This company's employees need a much better money manager.

Here is an example from my own essay: Critical regionalism became widely theorized **in** the 1980s **by** several architectural theorists as a means of realizing, embodying, and building **upon** a community's political consciousness **through** local spaces.

- Who is kicking whom? Who is the actor? Architectural theorists
- What is the action/kicking? Theorizing (Redundant)

Revised sentence: In the 1980s, several architectural theorists developed critical regionalism as a way of realizing, embodying, and building upon a community's political consciousness in local spaces.

Exercise: Now It's Your Turn

Use the paramedic method to help this sentence:

It is widely known that the engineers at Sandia Labs have become active participants in the Search and Rescue operations in most years.

What is your revised sentence?

VIDEO RESOURCES

Mayor Jim Gray on Using Writing to Solve Problems

Mayor Jim Gray was elected Mayor of Lexington in 2001. Mayor Gray is also the former CEO of Gray Construction, which in a successful international construction firm. As part of his daily work in the Mayor's office, Mayor Gray uses writing in a number of ways. In this interview clip, he talks about how he uses the writing process as part of his regular problem-solving process as Mayor.

(https://vimeo.com/103940590)

Stella Parks on Her Writing Process

Stella Parks is a pastry chef (formerly the head pastry chef at Table 310 in Lexington), a blogger, and a cookbook author. Stella is a native of Louisville, but she has lived all over the world. As part of her daily work as a pastry chef and blogger, Stella uses rhetorical theory and writing in a number of ways. In this interview clip, she describes her own writing process.

(https://vimeo.com/103940582)

Stella Parks on Writing Proposals

Here Stella Parks talks about her own process of writing a proposal for her new cookbook.

(https://vimeo.com/103938666)

Griffin VanMeter on Using Writing as a Creative Process

Griffin VanMeter is co-owner of the branding firm **Bullhorn Creative** *(http://bullhorncreative.com),* as well as the creative director of Kentucky for Kentucky (www.kentuckyforkentucky.com). Griffin grew up in Lexington, though he has also lived and worked around the country. As part of his creative work, Griffin uses writing and rhetorical theory in many ways. In this interview clip, he describes how he uses writing as part of his creative process.

(https://vimeo.com/104015855)

NOTES

1 Richard Lanham, *Revising Prose* (Boston: Pearson Longman, 2006).

Reading and Responding

5 Chapter

CAN YOU THINK IN QUESTIONS?

Have you ever played the game Twenty Questions? Someone chooses an object—often one that people can see just by looking around—and has the other people guess what it is. Usually the other people try to narrow down what the object is, asking "yes"-or-"no" questions:

- Is it cold?
- Can you kick it?
- Is it alive?
- Can I touch it from where I'm sitting?

We tend to call these "narrowing questions" because they help us to eliminate a whole range of things that may fall into giant categories. For example, if I say that the object is indeed cold, then you know it's probably not the fireplace with a roaring fire. It's also not the space heater across the room.

Imagine if I didn't try to ask such narrowing questions. Instead, I just started guessing at random: *"Is it that eraser there? Is it my glasses? Is it Dmitri?"* Everyone playing would look at me in disbelief. Why? It's because we use questions that are not too broad and not too narrow to get a better sense of what the answer is. Without having questions, we'd have way too many possibilities to consider.

In other words, questions help us limit the information we have at hand.

Reading is the same way. Questions can help guide us through the reading process because they offer projections for how a text might turn out. Reading requires us to be *active* in thinking through its information. Reading passively will only let you look at the information, but reading by inquiry will let you absorb it and think critically about it. You'll get a lot more out of your time spent reading that way.

Think of reading as a conversation between two people. One person can't speak alone, or she'd just be talking to herself. Reading is **dialogic**, which means that it depends on the co-creation of at least two people to make its meaning. An author may write words, but they don't mean anything until a reader reads them. Between every two sentences is a gap that you have to fill in. Usually we do this intuitively—without thinking. But at other times we have to actively draw and imagine connections between an author's ideas.

Author's Statement #1 ____________ (gap) ____________ Author's Statement #2

Look at the gaps between the sentences in these initial statements from Wendell Berry's essay, "Why I Am Not Going to Buy a Computer":

> *Like almost everybody else, I am hooked to the energy corporations, which I do not admire. I hope to become less hooked to them. In my work, I try to be as little hooked to them as possible. As a farmer, I do almost all of my work with horses. As a writer, I work with a pencil or a pen and a piece of paper.*[1]

First he's talking about energy corporations, and then he suddenly mentions that he works with horses. There's not direct link, but as inquiring readers, we supply the connection, since we know that the difference between big businesses and agrarian life will lead to his stance about technology which is given to us in the title. Some connections will be made for us by the author, but often we have to actively pursue the text's meaning.

So reading is more than just letting the words pass under your eyes. It means engaging in a dialogue with the author by seeing if her words make sense to you, if you agree with them, if you can follow their meaning from one section to the next, if they create a new idea in your mind that wasn't there before, or if they repeat or emphasize things you've learned already.

Not only will asking questions help you complete the conversation with the author through the text, but it will also give you a great way to take notes, especially if you write down the questions and jot down possible answers as you go.

READING REPORTS AND EDITORIALS

Choosing the kind of questions you ask while reading is just as important as choosing to ask questions in the first place. Some genres that you read encourage certain kinds of questions, and you're probably already familiar with some of these. Tweets respond well to the question, *"What's happening?"* Facebook status updates explicitly ask the question in their dialogue box, *"What's on your mind?"* which pushes people to compose a wide range of posts. Most texts you read or listen to can be thought of in terms of responding to a specific question or set of questions.

Newspaper reports are often written in response to the formula questions *"Who? What? When? Where? Why?"* By asking those questions about informational reports, you can easily follow the material presented. Facts tend to drive these articles, leading the reader along a path of causation in which the answers to those questions form a narrative that makes sense.

Reports may be full of opinions, but they tend not to be the main focus. Instead, we often see opinions in reports in place of facts. So asking the questions "Who? What? When? Where? Why?" means not only getting answers to those questions from the report, but also trying to figure out other answers to them that the report may not include.

Reports usually have a clear structure, explain their method, and present information as facts. They often marshal statistics, eyewitness accounts, and logical methods to arrive at some kind of cut-and-dried conclusion. In "Program Targeting Fraternities, Sororities Has Little To No Effect On Alcohol Use," we see a report that has a more empirical basis to it, the kind you will often find in the sciences (find it here: "Program Targeting Fraternities, Sororities Has Little To No Effect On Alcohol Use." DATA: The Brown University Digest of Addiction Theory & Application 18.12 (1999): 6. Academic Search Premier. Web. 15 Aug. 2013. Read it and decide what about it makes it seem "factual."

On the other hand, editorials (including the one we'll look at below) often respond better to the two-part question, *"What's wrong, and what do I think about it?"* since they take issue with something previously said. Usually engaging with hot topics, editorials are full of emotion as opposed to full of facts. They may provide facts to clarify a problem in the original viewpoint they are disagreeing with, but it is the emotions that organize their information. In addition to the question above, you might ask, *"How does the author feel about X and why?"*

Be aware that the questions you choose to ask are going to determine the kind of information you get. So you might ask editorials questions about their exigency (*What are the authors writing in response to?*), standpoint (*What's the angle on the topic?*), reasoning (*Why do the authors think they're right?*), or credentials (*why should you believe them?*). You can also search a text for questions that are a bit more distant from its primary purpose, such as *"What about this article made the newspaper want to publish it?" "How is this editorial's point similar or different from other arguments in the debate?" "What can I learn about stylistic technique by looking at this editorial as a good (or bad) model of writing?"* These questions develop a specific aim into your reading.

Consider the following **editorial by Brooke Myers written for the University of Oklahoma's student newspaper, the OU Daily, on Feb. 3, 2010** *(http://www.oudaily.com/opinion/column-fraternities-cleaning-duties-rush-and-the-quest-for-brotherhood/article_fef503e7-a3fb-550b-9a40-56221c4545f4.html).* As you go through, jot down questions you have, from the very first paragraph down through the last. And don't rush! Remember that quality inquiry comes from a sustained and focused reading of a text.

FRATERNITIES, CLEANING DUTIES, RUSH, AND THE QUEST FOR BROTHERHOOD

There is no bond like the bond of brothers. It is a bond made infallible. The word "brothers" connotes the dripping, sweet glue that binds men, young and old, together for life.

Together. For. Life.

Brotherhood has no limit, it has no age. It is an unconditionally bound relationship.

Let this be the eulogy to your days as an only child in this great, big world, dear newly-initiated fraternity boys.

Look, men. Look at the days that have passed. Remember them well, those preliminaries. Remember the beginning, when many of you were taken out to fancy dinners. Remember the delicious free meals and the kind smiles that were flashed at you. Remember the petty conversation that may or may not have reeled you in to your respective house. Remember being a chosen one, how good it must have felt to know you were wanted.

The Sorting Hat had done its bit, and you became another component in the category you apparently fit well. Some of you were jocks, some of you were cowboys, some of you were smart, some of you were stupid, some of you were attractive, some of you spent money in great amounts, some of you wore Polo well, some of you had impressive fluid capacities, some of you were good at getting girls, some of you really knew how to party. You all truly had your uses for your respective houses.

As new initiates, we can look back together on all the things you've had to do to make yourself worthy of admission. It has been a brutal trek, but you have made it.

When you were just wee little monkeys (we can't call you pledges, since it's illegal, and I rather like the name "monkey"—it has a cute little dehumanizing effect), you were made to do awful things. But those awful things were intelligently and symbolically designed so you would feel brotherhood seeping from your pores as you sweat while doing your duties, such as cleaning up the after-party leftovers in a fraternity bathroom while dressed in a nice suit. You were asked to be humiliated for a noble cause, making right-angle turns and speaking to no one but your pledge brothers for a week. You were deemed house-keeper at any hour of the morning, be it 2 or 5 a.m.

Boys, you were disrespected by the guys you would call your brothers, and you trusted they had dignified intentions; they trashed the house and yelled in your face while you were made to clean it all up. Your older brothers taught you lessons in presentation, forcing you to wear a suit everyday for a week without washing it. You were stripped of your identity, given no choices, brought down to a lowly level. You memorized names and numbers and more names and "star facts" about all the members of your respective houses. You were sleep deprived, abused, belittled, humiliated and asked to love it.

I envy you, though, for all that you gained from the experience. You formed a brotherhood with those experiencing the same struggles you were. You have learned, from deprivation, to appreciate so fully the honor bestowed upon you to be a part of something so much greater than yourself. You are one of them now.

You have truly accomplished something in life. You can now proudly display those prized Greek letters. You have been given a number—and it represents you! You are now four digits that say one thing: "I made it, I'm somebody!" And you are. You are somebody.

You've come so far. Your traits have blossomed. In fact, you might have added some you didn't have in the first place.

Many of you have truly been changed. And it's only the beginning.

Maybe, if you didn't once sport Polo and Sperrys, you do now. Maybe, if you didn't have long, side-swept hair under a backwards cap, you do now. Maybe, if you didn't wear cowboy boots originally, you do now. Maybe, if you didn't dip before, you do now. Maybe if you couldn't do the frat snap, you can now. Maybe, if khaki wasn't your color, it is now. Maybe, if getting chicks in bed wasn't on your agenda before, it is now. Maybe, if you didn't drink then, you do now. Maybe, if you cared more about school work than Greek letters, you don't now.

Maybe, if you were different from everyone else, you're not now.

Congratulations, guys, you're a whole lot like everybody else![2]

How do you respond to this article? Does it make you want to kick your desk? Stand up and applaud? Laugh violently? It made many students at the University of Oklahoma submit editorial responses to Myers that were published. Some attacked it for its strong, parodying tone, while others rushed to its defense with facts and evidence. Others got defensive and argued for the longstanding tradition of fraternities. Why did the article inflame so many different responses, and how do we deal with those responses?

Once again, questions come to the rescue. You can diffuse your own outrage or delight (or possibly indifference) to this article by approaching it with a series of questions to help you understand it. Now that you've read it, go back over the questions you jotted in the margin. Did they help you understand the article? How can you revise the questions you asked to better analyze the text? As you develop your own style of reading and questioning, you should get better and better at revising and adapting the kinds of questions you ask.

THE "SO WHAT?" QUESTION

Just because a text lends itself to certain kinds of questions for its production does not mean that other questions can't be used to understand it in different and possibly better ways (in fact, both editorials and reports often benefit from flipping the questions we ask about them, since they usually don't fall neatly into the categories of "facts" or "opinions"). For example, one of the most important questions in academic writing, *"So what?"*—which deals with the significance of a text—can be applied to any of the genres mentioned above. Think about a newspaper story—how would you read the story differently looking for the answer to "So what?" Or Facebook status updates—how many will you have to sift through to find a satisfying answer to that question? Choosing different questions to ask texts is encouraged by the textual genre itself, but also gives the reader latitude to search for what is important to him/her while reading.

This leads to another point about the kind of answers you'll find after asking these questions. The answers will change based on your own views and the context in which the text is published. Whether or not you care or know about a topic will influence your own judgment, but you also want to be aware that others may have very different cares or very different knowledge from you. So when reading a text through certain questions, you want to keep in mind both your own standpoint and what you know about others' standpoints. You may be a member of the main audience for some texts but part of a very peripheral one for others. Have you ever read a Facebook status or a tweet that didn't even make sense to you? The intended audiences of a text come from where and when it is published, and this shapes both its content and how we should read it.

Exercise: Different Questions for Different Genres

Find a newspaper editorial and ask the questions you would normally ask for a newspaper story. Write a paragraph about whether you were able to get answers to your questions and what that says about the genre of editorial. After you do that, come up with the questions you'd ask to understand the following newspaper genres:

- Comics
- Advertisements
- Weather forecast
- Dear Abby columns
- Classifieds

Put your questions together in a chart and see if you can find any patterns, such as commonly asked questions, among the newspaper genres.

How Do Emotions Fit In?

As Myers's extreme example of polemical rhetoric demonstrated, even the most objective reading is not free of emotions, since every stance taken by the author adopts a position that we may not agree with, that we may put up barriers against, or that we may agree with and accept happily. You should try to discover what emotions are trying to be provoked by the text and use your own emotive responses as a guide to making sense of the author's goals. Remember to maintain a grasp on the text's claims at the same time—what its argument and reasons are. We must combine emotional awareness with logic to imagine others' responses to a text, to gauge its rhetorical power.

Our position as an ethical audience leads to what rhetorician Peter Elbow calls "The Believing Game," in which we tacitly accept what an author says as true until we have read his/her rationale.[3] The notoriously divided Congress at present in Washington, D.C., might better reach compromise if they adopted this kind of reading and listening posture. Imagine a Republican arguing before the House for the benefits of closing abortion clinics and having all the Democrats present with rapt attention, nodding as they try to understand the speaker's position as if they were him/her. Or suppose the compromise that might be more quickly reached if the benefits of a lesser dependence on coal were jotted down while a Democrat spoke, rather than a list of arguments against it. Critical but ethical reading begins by acknowledging the values of the speaker's stance, not by simply rejecting them.

We've looked at two examples of "factual-based reports" and "opinion-based editorials" but as noted, the distinctions are not always so clear in most examples. Many authors will have heavily-slanted facts or strongly neutral opinions. Although some articles come close to relying exclusively on one or the other, such cases are rare and most examples will mix the two. Don't be fooled by the genre it claims to be part of, either. A report isn't any more true just because it claims to be news.

Exercise: Fact or Opinion?

Read the following report from the respected journal *The Chronicle of Higher Education* (Leo Reisberg, "Fraternities in Decline," *The Chronicle of Higher Education 46,* no. 18 (2000).

Which of the two genres does it seem to lean more toward? Why do you say that? Which sections come across as factual? Which as opinion? Identify parts that seem as though they're being passed off as facts but are really weighted opinions. Are there opinion sections that seem grounded in objective statements? Based on these questions, compose a paragraph in which you analyze the article for its logic, how it thinks. Remember to use your own questions that you've found useful to guide your reading and your evaluation.

Exercise: A Report and a Reply

Imagine you had to turn Myer's editorial into a report. How would you do it? Write up a plan with at least five steps for how you would go about making changes. Then do some research and write your own report based on Ms. Myer's topic. Finally, read back over both documents and make a list of the major differences you see.

Do you agree with Myer's view of the Greek system? Write an editorial in response to hers. What questions will guide your reply? How will they help structure your points? Make an outline of what you will say, using questions as your major headings. Now, write the reply. Finally, look at the two documents. What similarities and differences do you find?

RESPONDING TO PEERS' WORK

"Your first point has a well-developed structure."

"That word didn't make sense because it's not specific enough."

"Your tone seemed a little aggressive."

"The poor grammar in the fourth paragraph obstructs your point."

If you've received feedback on a paper, you're probably familiar with the language of critically evaluating essays, but it's not always clear how to provide that kind of feedback. In this section, we'll offer some skills that you can use to edit your peers' work.

Remember that making comments on your peers' work helps them with the process of writing. The reason papers written the night before the assignment's due don't turn out well is because good writing takes time and collaboration to develop. The more views and opinions you can get on your writing, the better you will know how your piece comes across to others and what you can change to improve it. It may take others saying that they didn't understand your fifth paragraph or that your best point was one you just tagged on near the end to get you to realize the real center and power of your work.

While we need others' help to make our writing stronger, most people are timid to make strong comments on others' work. Don't think that you're grading your peers, but that you're making their papers better by helping them improve their weaker parts. To do that, you'll want to give comments that are as specific as possible. Don't just say *"your words are vague"*—say which words in which paragraph are vague, why they're vague, and how your peer can make them less vague. Always identify the paragraph and sentence that your impression comes from. Always give a brief rationale of *why* that part is weak. Always offer suggestions for how to solve the problem.

Students and teachers alike are not always clear when they give feedback. So if a sentence seems awkward to you, don't just write "awk" in the margin—explain why it is awkward and how it can be corrected. If the tense is off, don't just jot "tense problems" but explain which tense the problematic area is in and why it should be changed to another. In short, clarity in your comments should mirror the kind of clarity you expect in your peers' writing. And of course, the clearer you write for your peers, the more precise and understandable your own essays will be.

Since the previous section offered some ways to use inquiry to actively read reports and editorials critically and ethically, let's see how you might apply those skills to editing your peers' work. Although the core aspects of reading and questioning will not change from published work to peers' work, your peers' work differs from published texts in that it is in process, and so your goal is to read, believing and questioning, with an end toward identifying areas that need more clarity, that lead the reader along, taking him or her through the gaps between sentences, from initial question to presented answer.

Now, just because your peer's work is still in an early stage of development does not mean that it will leave a single trail of breadcrumbs for the reader to follow—in fact, more advanced works often make the reader do more of the work or are open enough to let the reader imagine several possible and equally valid interpretations—but in general the connections between each point should be identifiable. Making sure the text is cogent and easy to follow is not always as simple as it sounds. It involves looking at the major ways language communicates.

The aspects of your peers' work that you'll want to look at may vary based on your instructor's specific requirements, but whatever those may be, thinking of the peer review at two different levels, global and local, will help you understand the kind of comments you're making on your peers' work.

- **First, think of *global,* or large-scale, changes to suggest.**

  Before worrying about grammar, look at the essay's thesis, argument, and main points. Does the argument make sense? Do all the points fit together? Does anything need to be added? Also ask about the order of the points. Does it follow some kind of arrangement model (for example, from most important to least important)? Once these broader and more important questions are answered, the smaller, stylistic components will gain relevance.

- **Next, identify *local,* sentence-level problems.**

  Grammar and style are best modified only after the ideas themselves are well established. Though second, these aspects can be critical in both the understanding and persuasiveness of an argument. Mechanical mistakes or stylistic dullness can easily overshadow the power of good ideas. Recall the editorial above: if its tone turned you off from the article, then you missed out on its ideas. Strong style that anyone will find compelling is one of the hallmarks of good writing.

For the first kind of remarks, think about what your peer's purpose is—and not just "to get a good grade on the essay." What is s/he trying to accomplish or say? Summarize the main point as succinctly as you can. In the rest of your questions, ask about how your peer unpacks the main point, asking for a logical organization and strong evidence.

For the second kind of remark, think about *how* your peer accomplishes that purpose. What does your peer do in the essay that successfully communicates his or her purpose? What about the style, tone, choice of examples, numbers of statistics, etc. are appropriate for following through with the main goal of the essay?

Use the checklist below to make sure you've covered the major parts of the essay.

Peer Review Checklist

Global

Thesis/ Argument	Is the point stated clearly?	Do all the points contribute to the main argument?	Is each point well supported with evidence?
Structure	Are the points clearly divided and without unnecessary overlap?	Are the points in the best possible order?	Should any paragraphs be moved or deleted? Should others be added?

Local

Content	Is the information about one main topic?	Is the language of the content as specific and narrow as possible?	Are transitions used to get from one piece of information to the next?
Style	Do paragraphs contain topic sentences, details to support the topic, and a concluding or transition sentence?	Are sentences formed with a variety of constructions, including different lengths, clauses, and subjects and verbs?	Is the style appropriate to the context of the assignment? How might tone or word choice be changed?
Grammar	Are tenses consistent?	Is punctuation correct? (Check especially for correct use of commas and correct combining of sentences.)	Is the third person maintained, except for the times when the information requires a shift?
Format	Are MLA conventions such as parenthetical citations used?	Is all information clearly designated from which source it has come?	Is the Works Cited page formatted correctly?

Remember that reading is a powerful act. Besides being a way of gaining information, reading may even reveal parts of a person much more intimate than those we glimpse in a face-to-face meeting. Part of the reason for this directness is due to the persuasive bent of rhetoric and part is due to the fact that a piece of writing establishes the values of the author, values that may be different from your own. Immediately, reading is a process of judging these values, negotiating them in ethical terms even while we are making sense of the lines of thought themselves. By practicing intelligent questioning of the texts we read, we can both have a better understanding of what they are trying to say and provide ways of helping others say what they want to say.

Exercise: Write a Peer Review Paragraph

Before writing anything, take your peer's essay and go through the process described above by looking first for global problems and then moving to local issues. After you have several ideas in your head, write a paragraph describing the main ways the essay could be improved. Be sure to write well yourself by referring to specific points in the text that can benefit from revision.

Exercise: Positive Feedback

Peer Review is not all about critique. One of the most delighting parts is to receive your own paper back with comments indicating how well you communicated. Identify three aspects of your peer's text that stand out as having been done well and write a paragraph explaining what they are and why they are done well.

Exercise: Rewriting Peers' Work

Rewriting your peers' work can give them ideas for better phrasing or pinpointing ideas. Try rewriting the following parts of a peer's paper:

- Come up with a new opening sentence for your peer's paper that gives a stronger introduction to the main points.
- Rewrite the thesis so that it is more concise.
- Change each opening and concluding sentence to each of the body paragraphs so that it introduces and wraps up the material in the paragraph and also transitions between paragraphs.
- In the concluding paragraph, come up with a new way of summarizing the main points without simply repeating them.
- Compose a new final sentence to the essay so that it ends with a strong sense of finality and closure.

NOTES

[1] Wendell Berry, "Why I Am Not Going to Buy a Computer," *What Are People For?: Essays* (Berkeley: Counterpoint Press, 2010), 170.

[2] Brooke Myers, "Fraternities, Cleaning Duties, Rush, and the Quest for Brotherhood," *OU Daily* (Norman, OK), Feb. 3, 2010.

[3] Peter Elbow, *Writing Without Teachers* (Oxford: Oxford University Press, 1986).

Chapter 6

Rhetorical Analysis

ANALYZING TEXTS

You've probably been asked to "analyze" texts, images, advertisements, or arguments. But what exactly does that mean? The term analysis comes from the Greek word *analusis*, which means to break apart. Whenever you analyze something, you are literally taking it apart in order to inspect its finer details.

When you analyze a song, for example, you might focus on its rhythm, harmony, lyrics, and vocals.

When you analyze a painting, you might break it into the colors, brush strokes, use of light, symmetry, and artistic genre.

Analyzing a rhetorical text requires the same kind of "breaking apart." What are the parts that comprise a rhetorical text? We can start with the elements involved in any rhetorical situation:

- Exigence
- Audience
- Author/ethos
- Text

The **exigence** is the situation, problem, or incident that caused the text to be written. If the text is a birthday card, then the exigence is obviously a birthday. However, if you're examining a card that says "I'm sorry" on the front of it, you might rightly imagine that someone did something that they now regret. In order to find the exigence of a text, ask yourself some of the following questions:

- Does this text respond to something that came before?
- Is there some kind of problem that this text is addressing?
- Is this part of a larger public conversation?
- Does this text fill some kind of need?

The **audience** for this particular text may or may not be immediately obvious to you. In fact, there may be several different audiences. In order to identify the audience(s), ask yourself:

- Who does the text seem to target through its language? (For example, is the language written at a basic level or in very specific jargon that only makes sense to experts?)
- Does the text use language that would make one kind of audience angry, happy, excited, or sad?
- Where does this text appear? Is there a certain kind of person who would usually read this magazine or website (or watch this TV program, listen to this podcast, etc.)?
- What kind of people are more likely to be invested in this topic?

The **author(s)** may be listed by name, or the text may be anonymous. Either way, you can decipher some important details about how the author tries to establish a certain kind of ethos in the text. The author/textual ethos is like the kind of reputation or persona interpreted by the audience.

Some authors want their texts to be read as highly professional or official. Other authors want their texts to be read as familiar, friendly, fun. One author may try to seem provocative and tough, while another author may want to seem approachable and non-threatening. In order to identify the authorial ethos, ask yourself the following questions:

- Is the author identified? If so, does she or he have a reputation already? Does this text fit with his or her reputation?
- Is the text identified with a certain brand (like a corporation or company)? Does this text seem to fit coherently with this brand's previous reputation?
- What kind of *persona* or *impression* does this text try to achieve or cultivate? Would you call this text cool, tough, friendly, informal, official, bureaucratic, etc? What makes you say so?

The **text** itself can be analyzed according to its style and genre. An advertisement online is quite different from a television ad for the same product. Likewise, a speech delivered to a room full of teenagers will use very different language, style, and visuals than a speech delivered to shareholders of a major corporation. When examining the text, ask yourself the following:

- What are the stylistic features of the text? Is the language vivid or very neutral?
- Are there images that accompany the text? What is the ratio of text to image, and what difference does that make?
- What is the context of this text? Is it part of a website? Is it a billboard on the highway? Is it a commercial shown during the Super Bowl? How does the context affect how an audience might read it?

Let's create a short rhetorical analysis of an advertisement by asking ourselves the questions above.

What seems to be the exigence for this text? For several years, Dove® has launched a public campaign to help fight the notion that women must be young and skinny in order to be considered beautiful. This advertisement is responding to a cultural notion that models (if not all women) should look a certain way (for example, "size 8"). In this way, the advertisement is part of a much larger public conversation happening about women's bodies and cultural standards of beauty.

Figure 1. Dove's Campaign for Real Beauty.[1] *(Image Courtesy of the Advertising Archives.)*

The audience for this particular advertisement is most likely women in their 20s and beyond. We can assume this because this ad (and others like it by Dove) appears in magazines and websites typically read by that demographic. The image included also shows an adult woman with whom adult women can identify.

The ethos cultivated by this advertisement is best described as positive and empowering,yet it also aims to create the impression of joining a debate. Rather than making direct statements, the language of the advertisement uses questions ("Fat?" "Fit?" "Does true beauty only squeeze into size 8?"). The image of a smiling, happy, strong woman who does not look like a "traditional" model seems to imply Dove's answer: beautiful bodies come in all sizes.

The text is simple, even sparse. There are few words, and those words are phrased as questions. The largest two words ("Fat?" "Fit?") are written as check boxes, as if the reader is supposed to "vote" on which term best describes the smiling, happy woman to the left. The negative connotation of the word "fat" forms a stark contrast with this joyful image. The fact that this advertisement appears in women's magazines, where there are ads with "traditional" models (extremely thin and young), may add to the overall sense that this text is part of a conversation.

ANALYZING ARGUMENTS

When you are analyzing arguments, you can include some additional elements that are present in any type of persuasive text. Consider the following questions involved in any argument:

What are the claims being made either implicitly or explicitly?

What types of evidence or support help make the case for this claim?

When looking for the **claims** in any argument, you might make a distinction between the main claim and any smaller claims also being implied or suggested. Try to decipher what it is that the author wants her or his audience to do, think, believe, change, or buy. What is the big point of this text?

Once you identify the claims, you want to then look for what kinds of evidence help to support these claims. In the language of rhetorical theory, what kinds of **proofs** are offered in support of the main claims?

Remember that rhetorical proofs (from the Greek term for "belief," *pistis*) are usually divided into three types:

Logos: the type of reasoning used in making an argument

Ethos: the type of persona communicated by the author

Pathos: the emotional, visceral, experiential elements of a message

Arguments do not simply use only one of these types of proof. Rather, every message has elements of reasoning (even if they're fallacious), a type of persona, and emotional or experiential elements.

Student Writing Example: Rhetorical Analysis by Natalie Watkins

WRD student Natalie Watkins wrote a rhetorical analysis of female integration into the armed forces. Natalie's analysis examines the ways that the army uses certain types of language and rhetorical appeals in order to communicate certain messages about female inclusiveness. As you read Natalie's analysis, notice how she tries to identify the different rhetorical elements involved in these messages.

RHETORICAL ANALYSIS OF SPACES

So far, we've talked about how to analyze texts and words. But it is also possible to analyze other kinds of rhetorical artifacts. Rhetoricians are fond of saying that "everything is an argument," which means that *everything has been designed by someone in order to accomplish a goal.*

Take a look around the room you're sitting in right now. What kinds of choices did the designers make? Are the windows big or small? Are the walls painted in vibrant colors, or are the walls neutral colors? How is the room set up? Are the chairs in rows, or are they arranged in a circle? What effects do all of these different choices make? You might say that these questions—and your answers—get us closer to a rhetorical analysis of a place. Even in the most mundane of spaces, like a classroom, someone has made specific choices in order to achieve certain effects on an "audience."

When you find yourself in a particular place, try asking the following questions in order to analyze the rhetorical choices that are used in that place:

- Is this a "man-made space" or a "natural space"?
- If this is a "natural space," are there man-made signs that help to "explain" this space?
- What are your eyes drawn to in this space? What are the areas that attract the most attention?
- Is this space designed for you to look, sit, move, or stand in particular ways?
- What colors are used in this space? What kind of effects do these colors have? How might entirely different colors affect you differently?
- Are you prevented from entering certain spaces or areas?
- What is the history of this place? Does the building or space have a sense of "timelessness," or does it seem to reveal a certain time period?
- How do you think the designers want visitors to feel in this space? What makes you say so?

Student Writing Example: Rhetorical Analysis of Memorial Hall by Patrick Weaver

WRD 110 student Patrick Weaver wrote a rhetorical analysis of Memorial Hall, one of our favorite spaces on the University of Kentucky campus. As you read Patrick's analysis, notice how he focuses on the physical details of the space, as well as its history. How does his analysis help you to see Memorial Hall in a new way?

A QUICK USER'S GUIDE TO RHETORICAL ANALYSIS OF TEXTS

When you are asked to create a rhetorical analysis, use this chart to help get started. You might find additional elements that are worth mentioning. If so, add those into your analysis! Remember that an analysis requires you to take apart a text in order to inspect its finer details.

Exigence	◎ Does this text respond to something that came before? ◎ Is there some kind of problem that this text is addressing? ◎ Is this part of a larger public conversation? ◎ Does this text fill some kind of need?	Give examples of how you came to this conclusion.
Audience	◎ Who does the text seem to target through its language? (For example, is the language written at a basic level or in very specific jargon that only makes sense to experts?) ◎ Does the text use language that would make one kind of audience angry, happy, excited, or sad? ◎ Where does this text appear? Is there a certain kind of person who would usually read this magazine or website (or watch this TV program, listen to this podcast, etc.)? ◎ What kinds of people are more likely to be invested in this topic?	Give examples of how you came to this conclusion.
Author	◎ Is the author identified? If so, does she or he have a reputation already? Does this text fit with his or her reputation? ◎ Is the text identified with a certain brand (like a corporation or company)? Does this text seem to fit coherently with this brand's previous reputation? ◎ What kind of persona or impression does this text try to achieve or cultivate? Would you call this text cool, tough, friendly, informal, official, bureaucratic, etc.? What makes you say so?	Give examples of how you came to this conclusion.

Text	◎ What are the stylistic features of the text? Is the language vivid or very neutral? ◎ Are there images that accompany the text? What is the ratio of text to image, and what difference does that make? ◎ What is the context of this text? Is it part of a website? Is it a billboard on the highway? Is it a commercial shown during the Super Bowl? How does the context affect how an audience might read it?	Give examples of how you came to this conclusion.
Claim	◎ What are the claims being made either implicitly or explicitly?	Give examples of how you came to this conclusion.
Evidence	◎ How does the author use logos? What types of reasoning are used in making the argument? Are there any fallacies or logically troubling aspects of his or her reasoning? ◎ How does the author communicate his or her persona? What kind of ethos does the text convey? ◎ What elements of pathos are involved? How does the text use emotional words or images in order to convey the message?	Give examples of how you came to this conclusion.

NOTES

1 Kylee Plummer, "PR Practice: Dove Campaign Presents 'Tick Box' Messages," *Dove Campaign for Real Beauty: The Real Truth about Beauty*, last modified June 9, 2013, http:// blogs.uoregon.edu/j350the-dovecampaignforrealbeauty/posts/.

Chapter 7

Argument

Argument seems like one of those things that you wouldn't necessarily need to learn. After all, even children know how to argue with their parents. We are all probably pretty good at being argumentative at certain times. But there is an important difference between what we mean by rhetorical argumentation and what might better be defined as "bickering." Rhetorical argumentation begins with what is called an exigency or an issue. There is some kind of stasis question that multiple interests are examining. At its most basic, argument has a claim and some kind of evidence to support that claim. Rhetorical argumentation always takes place in some kind of rhetorical situation, with particular audiences and constraints.

When students begin to learn rhetoric, the first question that many of them ask is "How can I be more persuasive?" Although no one can guarantee how to teach another person to always be persuasive, this chapter will look at two important aspects of rhetorical argumentation. First, we will look at the elements of argumentation in order to help you analyze and discuss the arguments of others. By understanding how effective arguments and ineffective arguments are constructed, you will have the power to better respond to the arguments of others. Second, by understanding how to construct a sound and compelling argument, you will increase your own ethos in a variety of situations where you want to persuade others. So, while the following information cannot guarantee that you will win every debate or argument, the information here can definitely increase your effectiveness as a rhetorician.

ARGUMENT BEGINS IN AGREEMENT

It may seem like a simple question to answer. Where does an argument begin? Take a minute to think about how you might respond to this question. You might say something like the following:

> An argument begins with a disagreement.
>
> An argument begins whenever two or more people have different ideas about the right way to think about something.
>
> An argument begins in moments of conflict.

All of these responses point to what we might call **stasis:** a point where we find ourselves standing at various distances from others who are concerned with the same question. For example, you may have an argument with a friend about whether or not it should be socially acceptable to illegally download music online. Maybe you feel like it is okay to do because it does not hurt anyone (like physically stealing something from someone). However, your friend has a different position on this question. Perhaps she feels that, over time, downloading music does hurt the artists who depend on music sales to make a living. You are both concerned with the same question—is it socially acceptable to illegally download music—but you both stand on different positions where the answer is concerned.

Your argument about downloading music may take all kinds of twists and turns. It may stay friendly, or it might turn serious. Maybe your friend gathers together some statistics about how much new or relatively unknown artists lose because of music downloads. She may even be so persuasive that you find yourself shifting your position. With enough evidence and compelling discussion, you might find yourself sharing her outlook on this question. Or, after arguing your own position, you may indeed be even more convinced that your position is correct. No matter what happens, however, you and your friend share a mutual concern for the same question: *Is illegally downloading music socially acceptable?*

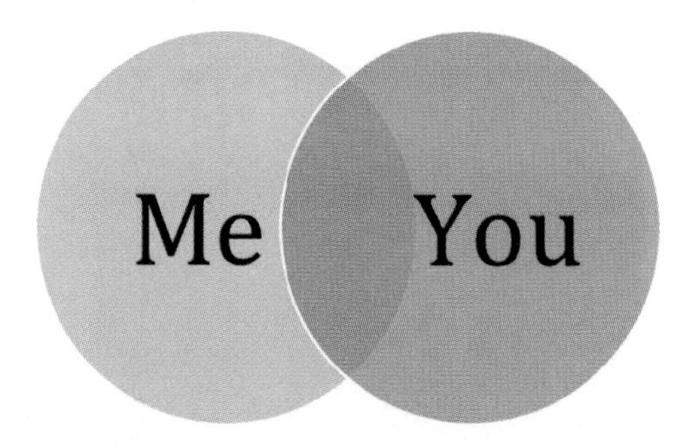

Ideally, when arguers share the same stasis questions, the deliberation and debates can be relatively productive. Perhaps you won't change people's minds on an issue, but you can contribute to new ways of thinking and an expanded knowledge base. Many rhetorical scholars believe that productive argument can be socially positive, even when there are no shifted positions. If you and your friend debate different issues relating to illegal music downloads, you might end up discussing such issues as how "stealing" should be defined, if breaking the law is ever socially acceptable, and even the ways that "property" should be defined when talking about digital material. This kind of debate could be an edifying conversation, regardless of whether or not your friend persuades you to stop downloading music.

> Beginning an argument with agreement might sound strange, but it's the only way to ensure that the process of argument is fruitful.

Sometimes, however, you might find yourselves in an argumentative situation that seems pretty unproductive. You have probably at least heard or witnessed these kinds of public arguments: people shouting at each other without listening, slogans that substitute for any real dialogue, even insults hurled back and forth. It might also seem as if two parties are "talking past" each other during an argument. One diagnosis of such unproductive argument is that the arguers are not truly in stasis. That is, they are not concerned with answers to the same questions.

Arguments about global warming show why it's important to begin an argument in agreement. Some writers have openly questioned whether global warming even exists, or whether climate scientists and journalists have falsely created this public crisis. One such global warming skeptic is Christopher Booker, who is a British writer known for questioning all kinds of scientific facts (including whether or not secondhand smoke is dangerous). Booker has written articles and books that debate the existence of climate change at all.

NEWS REVIEW

CHRISTOPHER BOOKER

"The laws of physics tell us that Al Gore's apocalypse could not come about"

Rise of sea levels is 'the greatest lie ever told'

Figure 1. An article from Christopher Booker in the *Daily Mail*.[1] (Daily Mail by Associated Newspapers UK Ltd.. Reproduced with permission of Associated Newspapers UK Ltd. in the format Educational/Instructional Program via Copyright Clearance Center.)

There are also a large number of scientists and experts who are concerned with a different question about global warming: what should be done about climate change? In 1997, the United Nations created the United Nations Framework Convention on Climate Change, which asked for all nations to reduce greenhouse gas emissions in their home countries. According to the language of this resolution (later called the Kyoto Protocol):

> *Recognizing that developed countries are principally responsible for the current high levels of GHG emissions in the atmosphere as a result of more than 150 years of industrial activity, the Protocol places a heavier burden on developed nations under the principle of "common but differentiated responsibilities."*[2]

The Kyoto Protocol resulted from significant debate about what should be done about greenhouse gas emissions. The debate included conversations about how more industrialized countries, such as the United States, would handle emissions differently than smaller and less industrialized countries.

One thing you might notice right away is that writers like Christopher Booker and those involved in authoring the Kyoto Protocol don't seem to share the same stasis question. That is, they are not trying to answer the same question.

Figure 2. United Nations' webpage on the Kyoto Protocol.[3]

Rhetoricians who agree with Booker are mostly interested in debating questions of *existence* and *fact*. They want to answer the question, "Is global warming real?" Rhetoricians involved in the Kyoto Protocol resolution are interested in debating questions of *proposal*. They want to answer the question, "What should be done about global warming?"

Imagine if these two different sets of rhetoricians were to engage in a debate. Booker and his friends may be unwilling to even entertain the question posed in the Kyoto Protocol—*What should be done about global warming?*—because they have doubts that global warming exists in the first place. Likewise, those who support actions like the Kyoto Protocol may be uninterested in engaging with Booker. After all, they feel like the question of global warming's existence has been well established and proven. If either side is unwilling to engage with the other's question, then they may end up "talking past each other" in a debate.

Exercise: Beginning an Argument in Agreement

Think of the last time you heard an argument that you found unproductive. What seemed to be the questions that each side was asking? Were they concerned with the same question, or were they talking past each other? Could the argument have been improved if the two sides had more explicitly agreed upon the question that they both wanted to answer?

ARGUMENT BEGINS WITH CLAIMS

In order to be productive, arguments must begin in some kind of agreement about the main question under debate. More technically, however, arguments usually begin with a **claim.** Rhetorical claims are what you put forward in any text (written, spoken, visual) in order to persuade an audience to believe or do something. Your claim may be explicitly stated, or it may be implicit.

An example of an explicitly stated claim:

- "You should vote for candidate X because she supports reforms in education!"

An example of an implicit claim:

- "3 out of 4 dentists recommend Toothies brand toothpaste." (Implicit claim: you should use Toothies toothpaste.)

Claims are sometimes also called your *thesis*, your *point*, your main idea. No matter what you call them, claims are the propositions or assertions you offer to your audience in any written, visual, or spoken text.

Claims may be minor, requiring relatively little development. For example, if you tell a friend that she should go see a movie with you instead of sitting alone at home, that is a claim. She may agree with you, but she might also disagree. Maybe she has too much homework, or maybe she simply wants to be alone.

On the other hand, claims may be major, requiring a great deal of support and development. If you want to convince your bank to lend you $10,000 to begin a small business, you will find yourself making a major claim about your business's feasibility, profitability, and sound business plan. The bank may agree with your claim that you are capable of successfully using $10,000 to launch a profitable business, but they may also disagree with you.

Whether minor or major, claims are almost always *debatable*. While you may find your claim 100% convincing and persuasive, there are people in your audience who might present reasonable objections to your claim. (Even your friend sitting alone could reasonably contest your claim that she should accompany you to the movies!)

Think of how many claims you have made or heard in the past 24 hours. If I could list all the claims I've made recently, they would include the following:

- *This winter is too cold!*
- *I should start making my own lunches in order to save money.*
- *The new Beck album isn't really all that great.*
- *Standardized tests in public schools don't help students learn more.*
- *We need to hire more employees at my workplace.*

Not all of these claims are controversial. It's hard to imagine someone trying to passionately debate me about whether or not making my own lunches really saves money. (However, my friend did recently send me an article that questioned how much money you can save by making your own lunches!) Some of the claims, of course, are very controversial. Standardized testing in public schools sparks impassioned responses from both detractors and supporters. Even music reviews can lead to animated debates between those who find an album great or terrible.

EVIDENCE

Few audiences simply accept claims without much evidence or support. Evidence is a crucial piece of every claim. If you make a claim without evidence, you are simply making a statement. Consider the differences between these scenes.

Scene One

Amelia: You shouldn't buy that car.

Rocky: Why not?

Amelia: You just shouldn't.

Scene Two

Amelia: You shouldn't buy that car.

Rocky: Why not?

Amelia: It gets horrible reviews online.

In the first scene, Amelia tells her friend Rocky that he shouldn't buy a certain car, but she gives him no real evidence that buying the car would be a mistake. By asking "Why not?" Rocky is asking Amelia to give him some kind of evidence that would make her claim sound reasonable to him. In other words, he wanted some evidence that he should take Amelia's claim seriously. She doesn't provide any evidence, however. Chances are low that he will simply take her word for it.

In the second scene, however, Amelia does provide Rocky with evidence. She tells him that the car he's hoping to buy gets horrible online reviews. This is a small piece of evidence in support of her claim, and it might not initially be enough to persuade Rocky against buying the car. However, it does offer some kind of evidence that Rocky should take Amelia's claim seriously.

Notice that by providing evidence, Amelia is not guaranteed to convince Rocky of her claims. Rocky may be extremely skeptical of anything that appears online. Or he might want to know more about where these reviews are found and who is writing them. Providing evidence is not a guarantee that your argument will be persuasive, but it might prevent listeners from dismissing you right away.

If you were to visualize most arguments, they would probably look something like this:

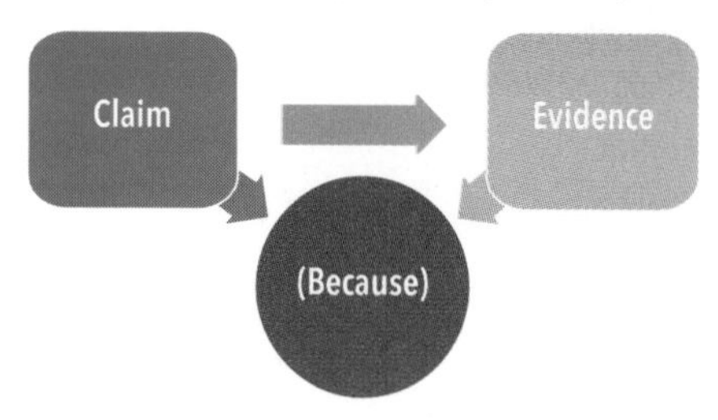

The word "because" links the claim and evidence, since the evidence is what explains any argument's claim.

For example, Amelia's second argument above would look something like this:

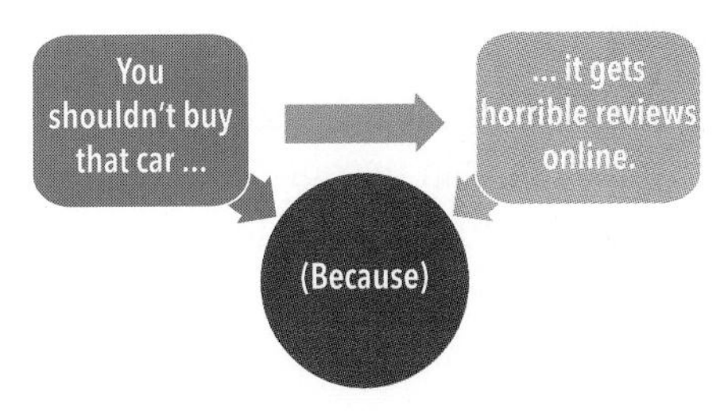

Even in instances of implicit claims, you can draw this argument diagram to see if evidence is actually being offered. Earlier, I gave an example of a common type of implicit claim: "3 out of 4 dentists recommend Toothies brand toothpaste." What would this implicit claim look like as a diagram?

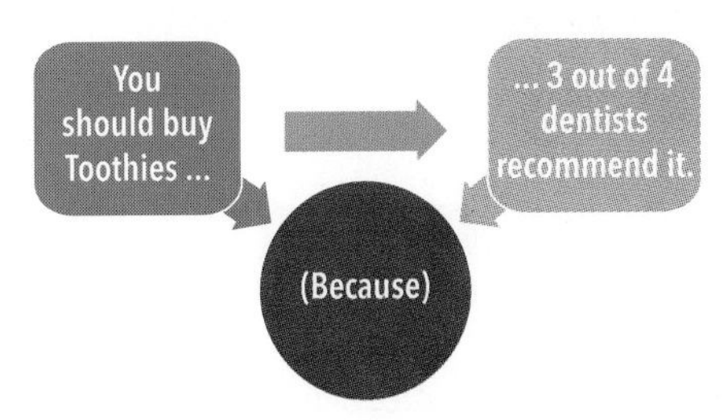

Many claims are more complex than the examples above, of course. In real world arguments, several pieces of evidence may be offered in support of a claim. For instance, the language of the Kyoto Protocol argues that developed nations should bear a greater portion of responsibilities for correcting climate and environmental damage. If we tried to diagram this argument, it might look something like this:

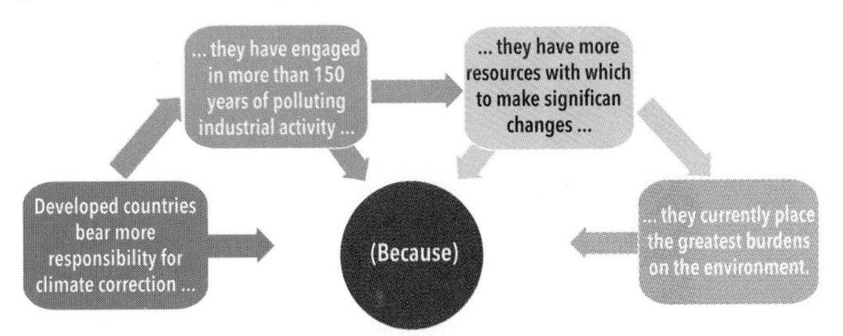

In order to check and see if you are providing evidence for your claims, try the following exercise. Look over your argument and use the "because" diagram above in order to draw your major claim and evidence. Do you offer a "because" in order to support your claims?

WARRANTS

Unfortunately, even if you offer the strongest possible evidence you can find in support of your claim, your audience may remain unconvinced. Why is this? Is this because your audience doesn't understand your argument? Are they just being stubborn? Chances are, when you have really made the strongest possible argument you can make and you *still* find opposition, you are experiencing someone with a different **warrant.**

Warrants are the third part of any argument. Claims and evidence are held together in a bond, and warrants are the glue that holds those claims and evidence together. Warrants are the underlying beliefs, opinions, ideas, and feelings that allow for your evidence to logically support your claim (or not).

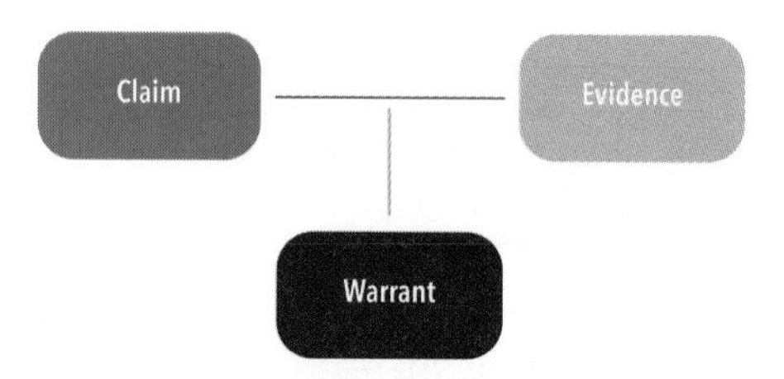

Warrants are usually unstated and unnoticed by everyone involved in an argument. In fact, you might not even think about your own warrants until you find yourself questioning whether certain claims and evidence really work together.

For example, in Amelia's claim that Rocky shouldn't buy a certain car because it gets horrible reviews online might not convince Rocky. He might argue that he doesn't believe online reviews. In fact, he finds them very unreliable and even shady. Amelia disagrees, however. She finds online reviews a very credible and reliable way to decide on products.

The argument between Amelia and Rocky might look something like this:

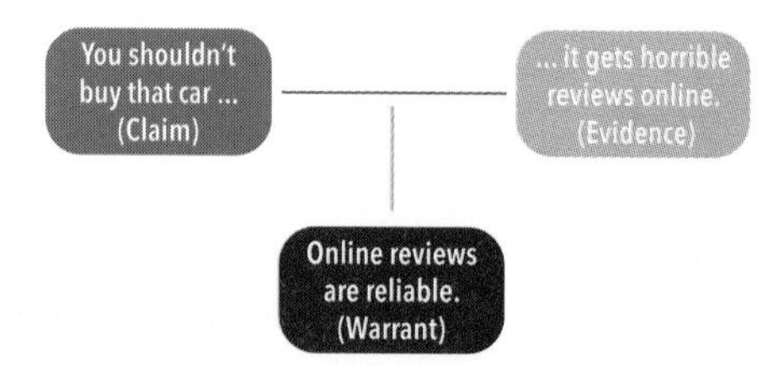

In order for Amelia's claim to be persuasive to Rocky or anyone else, they must share Amelia's warrant that online reviews are reliable. However, Rocky does not share this warrant. He feels that online reviews are unreliable. Therefore, the warrant does not "glue" Amelia's claim and evidence together in a way that persuades him.

Warrants can be described as cultural, social, religious, and ideological beliefs that allow us to believe certain things and not others. If your warrant is radically different from your audience, then you may find that making a claim and providing persuasive evidence is very difficult. Consider, for example, those who make arguments with others based on the Christian Bible.

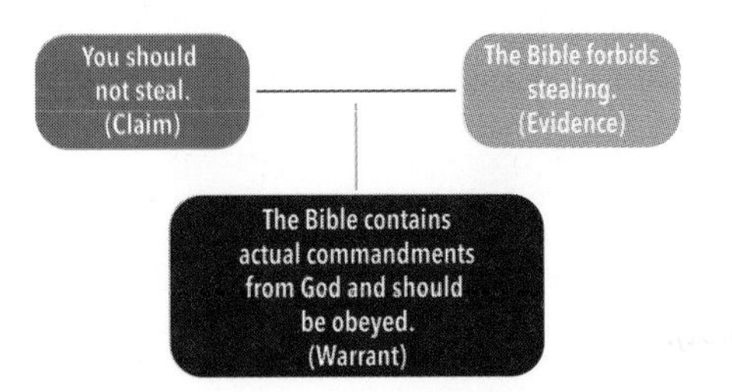

If both parties share the same warrant that the Christian Bible is true and infallible, then this evidence may indeed be persuasive enough to convince the other party. However, if one party in the argument is not Christian, then the warrant may not be shared.

If I do not believe that the Bible contains actual commandments from God and therefore should be obeyed, then the evidence that the Bible forbids stealing will not necessarily persuade me that stealing should be avoided.

On the other hand, you might be able to offer me different kinds of evidence for this same claim. Perhaps you want to draw on my sensitivity to others and my sense of compassion. Therefore, you might change your argument to look more like this:

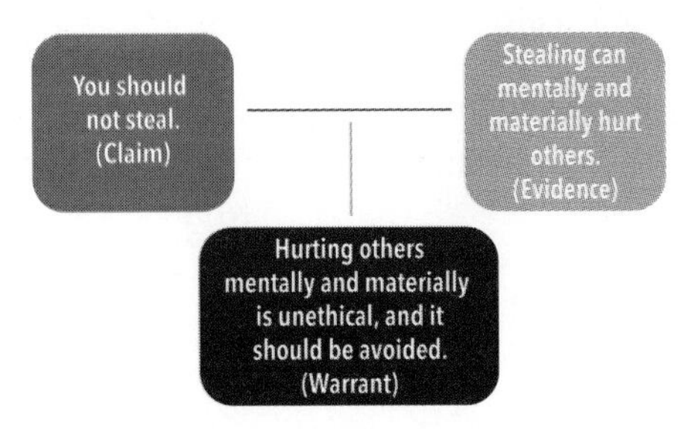

This different warrant may be one that I do indeed share with the arguer. In this case, I would find the evidence offered in support of the claim ("stealing can mentally and materially hurt others") to be very compelling.

LOGICAL ARGUMENTATION

Another way to think about arguments is to think about them in terms of simplified claims that logicians and philosophers call "statements." While rhetorical claims are concerned primarily with persuasion, statements are concerned with facts. A **statement** is any simple sentence that is either true or false. Therefore, a logical statement *must* be a declarative sentence or part of a sentence which could stand alone as a declarative sentence. Here are a few examples:

- The sky is blue.
- You are likely to need an umbrella today.
- Pizza is not a vegetable.
- No students ever cheat on their papers.
- Plenty of rest helps you retain information better.

Each of these sentences can at least in theory be determined to be true or false. The first two are true or false depending on the time of day and weather conditions. The third sentence is false based on the definitions of "pizza" and "vegetable." The fourth sentence can be determined to be false by just one counterexample (i.e. one student who cheats one time). The fifth sentence can be determined to be true or false by scientific study.

However, there are some statements which do not have this capacity to be determined true or false at all. These include questions, commands, suggestions, opinions, and so forth. These are certainly grammatical sentences, but a sentence is not always a statement. These are examples of sentences which are not statements:

- Do you know the way to Memorial Hall? (question)
- Be quiet and put away your phones. (command)
- I think you need a haircut. (suggestion)
- I like tacos more than burgers. (opinion)
- Let's go to the basketball game! (proposal)

Sometimes the use of questions, proposals, and commands can be rhetorically persuasive in terms of ethos and pathos, but these kinds of sentences are counter-productive if you want to focus on logos. Strictly speaking, then, an **argument** is a series of statements where one or more statements called **premises** provide support for usually one statement called the **conclusion.**

Let us be very clear: In an essay, we use the word "conclusion" to refer to the final paragraph or the ending of the essay. However, an **argumentative conclusion** is somewhat different: The conclusion in an argument is the statement that you are trying to prove to be true with the use of the other statements. The conclusion is the whole point of the argument, the reason you are offering an argument in the first place, or the thing you want to persuade other people to believe. Usually we use signal phrases to indicate that a certain sentence contains our argumentative conclusion. These include words like **therefore, thus, consequently, accordingly, hence, it follows that, it implies that, so** and more. Be careful, though: Many writers use these words without any argumentative purpose. Thus, just seeing a signal phrase does not mean that you have always identified an argumentative conclusion–it can mean that the writer just likes to sound important, or that they are using the word in grammatically correct but non-logical senses.

Similarly, we have many different meanings for the words "premise" and "premises," including the starting point in the plot to a story (the "premise" of a movie) and the grounds of a building (the "premises" of the University). However, **argumentative premises** are again somewhat different: The premises in an argument are the statements that you are using to prove your conclusion. Premises include both warrants and evidence, and are the reasons why the conclusion is supposed to be true. With premises the rhetor provides justification for the claim they are trying to assert is true. If the conclusion is the "what" behind what the rhetor is saying, the premises are the "why." Premises also have signal words, including **since, because, for, given that, for the reason that, as indicated by** and others. Again, however, you should remember that these can be used without any care for their logical significance.

Arguments in ordinary language are not always clear and logical exactly because they are in ordinary language. When we speak to our family and friends we make assumptions about what they know, believe, and value, and so we leave many of the claims in our arguments unstated. We also make complex arguments, where the point of one series of reasons leads to another point, and so on. These kinds of arguments can be translated into clear, categorical language that reveals their structure and makes it easier to spot logical flaws or fallacies. Translating arguments into simpler language also helps us to critically analyze the arguments we receive from others.

There are several ways to criticize an argument, even without taking a logic class which requires you to know the difference between *enthymemes* (arguments which leave some claims unstated) and *sorites* (arguments which link together). The two primary methods are analyzing the content of the premises and analyzing the structure of the premises. Consider the following examples:

- All ducks are mortal.
- Socrates is a duck.
- Therefore Socrates is mortal.

In this example we have a true conclusion (Socrates is mortal) based on a true premise (ducks are mortal) and a false premise (Socrates is a duck). In responding to this argument, we can say that it is a bad argument because one of the premises is false—the facts are different, Socrates happens to be a human. Compare this example with the next:

- All humans are mortal.
- Socrates is a mortal.
- Therefore Socrates is a human.

In this example we have true premises (All humans are mortal; Socrates is a mortal) and a true conclusion (Socrates is a human). However, there is still something wrong with this argument: Just because those premises are true does not prove anything about the conclusion. Sure, all humans are mortal, but so are butterflies and trees and millions of other things. Just because Socrates is mortal, this does not prove that he has to be human. Thus, there is a structural problem with this argument. In responding to this argument, we can say that it is a bad argument not because of a false premise, but because the conclusion does not *follow from* the premises.

This is a structural or *formal* logical problem, and it is a bit trickier than simple truth or falsity, because this is a matter of **validity** or it is a matter of **strength.** Again, in ordinary language, 'valid' (and 'strong') has a number of different meanings, including legitimate, right, acceptable, and so forth. However, **argumentative validity and strength** apply only to the structure of the argument, and whether the way the premises fit together actually allows them to build the conclusion. This means that a single statement cannot be valid or invalid, nor can just any group of statements, nor can just any argument. If you want to sound like you know what you are talking about, do not use the words 'valid' or 'strong' to simply describe a legitimate opinion or a good point.

Logicians divide arguments into two groups: deductive and inductive. **Deductive arguments** are determined to be **valid or invalid** on the basis of the structure and relations between the premises and conclusion. Deductive arguments include: arguments based on definitions, arguments based on mathematical proofs like in geometry, and syllogisms. Syllogisms are typically arguments with two premises and one conclusion only, and include categorical syllogisms, disjunctive syllogisms, and hypothetical syllogisms.

A categorical syllogism relies on the structure of categories and will usually have statements like "All cats are mammals" or "Some fish are not freshwater animals." By contrast, disjunctive syllogisms rely on the structure of 'or' statements like "Either he knows French or he knows Spanish," and hypothetical syllogisms rely on the structure of "if/then" statements like "If you do not turn in your excuse, then you cannot make up the work." The quality of deductive arguments depends on the form or shape of the arguments; which piece is where in relation to the others, and so forth.

By contrast **inductive arguments** are determined to be **strong or weak** on the basis of the structure and quantity of the premises in relation to the conclusion. These are more familiar arguments because they provide the basis for much of our daily lives. Inductive arguments include: predictions, generalizations, causal inferences, and analogies, as well as arguments based on signs or the authority of the source.

A few examples of these include weather forecasting (prediction), statistics based on surveys (generalizations), diagnoses of the reason you are sick (causal inferences), using the behavior of one city to support policy change in another similar city (analogies), concluding that there is no class today because there was an email or a literal sign in the classroom (argument by signs), and claiming that the Big Bang occurred a certain way on the basis of Stephen Hawking's work (argument from authority). The quality of inductive arguments depends on the form or shape of the argument as well as the sheer quantity of evidence. More examples cannot improve a deductive argument, but an inductive argument can often be improved by adding more and more examples, provided they are of otherwise good quality.

FALLACIES

A flaw in an argument is called a **fallacy** and may be formal or informal. Deductive arguments with structural or formal flaws are invalid; inductive arguments with structural flaws are weak. However, both deductive and inductive arguments can have informal flaws as well, and these do *not* make the arguments invalid or weak; instead, a deductive argument with an informal fallacy is unsound, while an inductive argument with an informal fallacy is uncogent. **Soundness** and **cogency** are evaluations of the overall quality of the argument, not just the structure, but also the truth of the premises, the truth of the conclusion, and the other considerations raised by informal problems.

Rhetorical fallacies tend to be fallacies of the informal type. These informal rhetorical fallacies are present in many arguments, though they may not be immediately obvious to the audience. Sometimes, the evidence offered in an argument may sound perfectly reasonable, at least until you really pay attention to what's being said. The value of recognizing a fallacy is that you can draw attention to the faulty logic being passed off as solid reasoning. At the same time, you can also work to consciously avoid using fallacies that might land you in trouble with a savvy audience. Here is a partial list of some common rhetorical fallacies.

Deductive Fallacies

Of Relevance

- **Ad Hominem:** Making a personal attack to distract one from the substantive issue.

  Example: *Candidate Smith can't lead the country because she has a terrible personality.*

- **Ad Populum:** Appealing to desires, fears, and prejudices of the audience to arouse mob mentality and distract from the issue.

  Example: *Real men use Bro body spray.*

- **Red Herring:** Misleading information designed to pull the audience off track from the main point under discussion.

  Example: *Yes, I am guilty of bank fraud. However, I am an excellent friend and spouse. Everyone who knows me knows that I am kind and generous.*

- **Poisoning the Well:** Planting accusatory or negative information in misleading questions or statements.

  Example: *Would you still vote for Candidate Yee if you knew she did not support government research into finding a cure for cancer?*

- **Straw Man:** Distorting the position of the opponent and then attacking the distortion instead of the opponent's actual position.

  Example: *The Senator has argued against prayer in public schools. Clearly, he is an atheist, and he is trying to make all our children atheists as well.*

Of Presumption

- **Begging the Question:** Using circular reasoning; creating the illusion that the premises support the conclusion by restating the premise in the conclusion.

  Example: *Organic food is healthy because it's good for you!*

- **Complex Question:** asking two or more questions simultaneously in the guise of a single question, thus generating a single answer to the multiple contained questions.

  Example: *When did you stop cheating on your exams?*

- **False Dichotomy:** Using a disjunctive statement ('either/or' statement) which obscures the possibility of more than two choices.

  Example: *This is America, and you can love it or leave it!*

Of Grammar

- **Amphiboly:** Using an ambiguous statement to draw faulty conclusions.

  Example: *Dr. Smith donated, along with his wife, about $200,000 at the charity event. The organization is grateful, but they have no place to keep Mrs. Smith and had to send her home.*

- **Equivocation:** Using a word or statement in two completely different ways in premise and conclusion.

  Example: *The U.S. Constitution says all men are equal, but women are not men. Therefore women are not equal to men.*

Inductive Fallacies

Of Weak Induction

- **Unqualified Authority:** Using authorities or witnesses who have little or no credibility with regards to the argument in question.

  Example: *Albert Einstein said God doesn't play dice with the universe. God must exist if that genius thinks so!*

- **Appeal to Ignorance:** Using the absence of evidence as the evidence of absence.

  Example: *No one I know has ever seen God. God must not exist.*

- **Hasty Generalization:** Drawing a conclusion from too few examples, or too little evidence.

  Example: *My Russian neighbor dislikes cats, so all Russians must dislike cats.*

- **Post Hoc, Ergo Propter Hoc:** "After this, therefore because of this." Assuming that because one event preceded another, it must have caused the other.

  Example: *When I wear my lucky socks, I get an A on my test.*

- **Slippery Slope:** Alleging a simplified chain reaction in complex situations where the proposed chain of reactions has little to no connection to the original cause.

  Example: *If we allow people to marry others of the same gender, then soon people will be asking to marry dogs and cats, corpses and children!*

Of False Analogy

- **Faulty analogy:** Using unlike terms in an analogy.

  Example: *Raising a baby is like raising a plant. If you can't even raise a plant, then you shouldn't have children.*

- **Composition:** Transferring the attributes of the parts to the whole.

  Example: *Those politicians are corrupt individuals. Therefore, our entire government system needs to be completely reformed.*

- **Division:** Transferring the attributes of the whole to the parts.

  Example: Car engines are very heavy things. I have to pick up an air filter and spark plugs today, how on earth do you expect me to bring those home on the bus?

You can find many more lists of fallacies by simply searching the web for rhetorical fallacies!

For more information, see Chapters 1 and 3 of *A Concise Introduction to Logic* (10th edition) by Patrick J. Hurley (Belmont: Thomsom Wadworth, 2008).

STAKEHOLDERS

Sometimes we mistakenly think of argument as a black-and-white debate between two sides: pro and con. We even talk about trying to see "both sides" in an argument. But, in actuality, this is not how arguments over public issues play out. Instead of "two sides," there are different kinds of **stakeholders** who are invested in or affected by the problem being discussed. Stakeholders are those who may participate in an argument with different perspectives.

Consider the public debates over whether or not adults should be allowed to carry concealed weapons on university campuses in Kentucky. Many different stakeholders have weighed in on this question, all bringing new perspectives to the conversation.

- The University of Kentucky has strictly enforced a "no gun" policy on campus.
- A local student group called "Students for Concealed Carry" has lobbied the University to allow concealed permit holders to carry guns on campus and even into classrooms.
- UK faculty members have declared their opposition to concealed handguns on campus.
- UK students who do not hold licenses may object to sharing classroom or dorm spaces with guns.
- Parents who send their children to UK hold various opinions on whether concealed weapons make everyone safer or put more people at risk.
- Dorm resident assistants and directors have indicated that the presence of guns would require them to "police" deadly weapons.
- The Kentucky Supreme Court ruled in 2012 that UK students and staff may keep licensed weapons in their cars on school property, though they may not carry them anywhere else on campus.
- The city of Lexington must also deal with potentially thousands of new guns flowing into Lexington, thanks to students who now live on campus with guns. How will these guns and their sale be regulated?
- National pro-gun organizations, such as the National Rifle Association, have been very vocal in their support for concealed guns nearly everywhere, including campuses.
- National gun control organizations, such as the Brady Campaign, have also been very vocal in their opposition to guns in schools and other public spaces.
- Several survivors of the Virginia Tech shooting massacre have spoken out against guns on campus. See the documentary *Living for 32*.

As you can see from this list, there are many different groups who are invested in or are affected by the question of whether or not guns should be allowed on UK's campus. Not all of these stakeholders can be easily categorized as "pro" or "con." Some stakeholders bring entirely different perspectives or issues to the table. For instance, the Kentucky Supreme Court has been an important voice in this debate through its interpretation of the law.

Likewise, dorm directors may not be opposed to guns per se, but they may not want the responsibility of "policing" guns among dorm residents. Likewise, faculty may not be opposed to gun ownership, but they may not feel comfortable with students carrying guns into the classroom.

How Do You Find Stakeholders?

Finding stakeholders is a process of finding all organizations, individuals, groups, professions, and people who would be affected by this topic or who have spoken up about this topic. List as many as you can think of, and then search for additional stakeholders. You might search the web for organizations or groups who have something to say on the matter. Are certain people or groups quoted in newspaper articles about this subject?

It might help to actually draw, or map, these stakeholders to get a sense of their relationships to one another. Furthermore, your map can also represent their potential impact in this debate. Which stakeholders have the loudest voices, so to speak?

If I was going to map the various stakeholders in the UK concealed guns debate, it might look something like the image below:

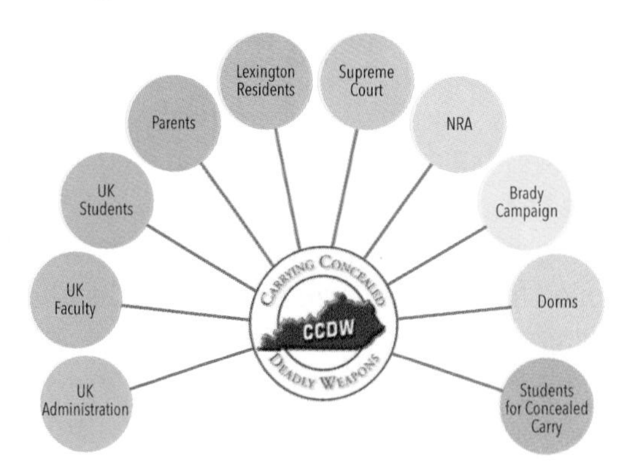

STAKEHOLDER ANALYSIS

After you identify all the stakeholders you possibly can, now it is time to engage in a **stakeholder analysis.** A stakeholder analysis is a genre quite common to businesses and organizations that are trying to identify those who may be allies or adversaries in advancing some argument. By knowing how to create stakeholder analyses, you will have an advantageous and useful skill for whatever career you enter. The following will lead you through the steps to conduct a stakeholder analysis:

1. For each stakeholder, list its characteristics. What does this group, organization, or person value the most? What are its priorities? What does this stakeholder most want? What are its fears? How will this situation impact them?
2. Next, list this stakeholder's rhetorical characteristics. What kinds of texts does it use to make its case? What type of language does it use? What types of rhetorical appeals are especially evident in their discourse?
3. Finally, analyze this stakeholder's main claims. What claims does this stakeholder make (about your chosen topic)? What types of evidence are offered in support of these claims? What kinds of warrants are "bridging" this argument for the stakeholder?
4. When you are finished with your analyses of all stakeholders, see if you can identify any points of agreement, disagreement, similarities, or patterns among the various stakeholders. Even among those who disagree, are there rhetorical similarities in their discourse or rhetoric? Do they rely on similar or different types of arguments? Are their warrants different, or are they generally shared?

NOTES

1 Tim Holmes, "Christopher Booker uses Times Atlas story for his own purposes," *The Carbon Brief,* last modified Sept 22, 2011, http://www.carbonbrief.org/blog/2011/09/does-booker-believe-in-listening-to-experts.

2 "Kyoto Protocol," United Nations Framework Convention on Climate Change, accessed June 16, 2014, https://unfccc.int/kyoto_protocol/items/2830.php.

3 Ibid.

Chapter 8

Using Evidence

The phrase "using evidence" sound like it belongs more in a courtroom or a crime scene than in a writing classroom. When we think of this word, a whole host of images might pop into our head:

- A police officer putting on gloves and picking up an object in an area cordoned off with yellow crime scene tape.
- A reporter squinting at a computer screen or shifting through stacks of paper in search of a piece of information for a story.
- A lawyer in a courtroom, holding up a weapon in a numbered bag and using it to prove his case to a jury.

Of course, when we hear **"evidence,"** we might also think of more traditional synonyms or definitions, such as an object or document that includes the following:

- Provides **proof** or **corroboration.**
- Functions as a form of **validation** or **authentication.**
- Acts as a vehicle of **persuasion.**
- Serves as **documentation.**
- Indicates **truth** or **accuracy.**

The item that the police officer finds at the crime scene; the information that the reporter uncovers; the weapon that the lawyer holds up to the jury in the courtroom: while each may be physically different, they all function in a similar way because they all serve as a kind of **evidence.**

It's kind of comforting to know that people like police officers, reporters and lawyers all have to find enough evidence to make their case or their argument. But it can be unsettling when we're asked to do it ourselves—especially in our writing.

As we discussed earlier in this textbook in the chapter on rhetoric, **artistic arguments**—arguments that use ethos, pathos, and logos in some form or fashion—allow us to use the credibility, emotional connection, and sensible structure that is largely our own invention to make our point.

But the kind of evidence that we're discussing in this chapter is the kind most often used in **non-artistic arguments**—in other words, it's the kind of evidence that isn't our invention. It's the kind of evidence that we have to seek out from other people and other reliable sources, the kind of evidence that changes from assignment to assignment and genre to genre. Sometimes that evidence might come from a personal interview or an article from an academic journal; other times, it could be an excerpt from a novel or a newspaper article.

Sometimes when we're asked to use evidence in a piece of writing, we have a visceral reaction to this request. If we don't have much experience doing research or incorporating evidence into our writing, it can be intimidating, even a little frightening. What, we think, is my word not enough? Where am I supposed to find this "evidence"? How will I know if it's any good? And what on earth am I supposed to do with it once I have it?

Why do we react this way? Because, first of all, it's natural to be a little nervous about writing something in a different format than we're used to, especially if it requires more steps and more rigorous analytical thinking than we may have used in the past.

In addition, writing is too often portrayed culturally as some kind of magical, mysterious, mystical process that some people are good at and some people aren't, and we often assume that if we don't know how to do it right away, then we must be the person who isn't good at it.

But as natural as the nervousness is, this perception of writing as some kind of magic is a myth. In truth, clear, effective writing is largely achieved by learning to implement a few rules and putting in some hard work. This, too, is the key to using evidence effectively.

In this chapter, we'll outline and explain four steps in this process, which will help you with the following:

- Determine which sources are best for your piece of writing.
- Understand what your sources are arguing.
- Incorporate those sources into your writing so that they bolster your claim.
- Revise your work so that your use of evidence is most effective.

STEP ONE—REVIEW: WHAT DO I HAVE?

Looking for sources can be both time-consuming and tedious—hours in the library sifting through the stacks for that perfect book or journal, hours on the Internet reading link after link, trying to find the best piece of relevant information. Although it's sometimes frustrating, doing research and doing it well does require the dedication of some time. And when you finally sit down to write that paper, you don't want to stare at a stack of books or articles or links and wonder, what do I have here? Before you pick up a pen or open that Word document, you need to have a good idea of what you have in your possession, so you have to **evaluate your sources.** Here are a few tips on how to do that:

> Be discriminating: Make sure you are using the best, most reliable and trustworthy sources possible.

Although Wikipedia might seem like a go-to for gathering some quick research on a topic, a crowd-sourced website that allows anyone to edit its entries isn't as trustworthy as, for example, a website run by a college or university or a government agency.

It's certainly better to get information about the resurgence in measles or the latest Ebola outbreak in Africa from, say, the National Institutes of Health, the nation's top medical research agency which is staffed and run by some of the world's best doctors, rather than from the Wikipedia entry for either of these topics, which could have been edited by anyone (and chances are, it isn't someone with a medical degree). For more on why choosing the most reliable sources is so important, you'll find a real-life example in Case Study 1 on pg. 163.

> If a source doesn't seem to work for you, the bibliography might.

It's frustrating to realize that sometimes a source that you've worked hard to find isn't the best source for you at all (see the box "The Pain of Tossing out Sources" on pg. 161). However, that doesn't mean that the source is completely useless. Before you toss that source, make sure you check its bibliography: many times, you'll find a host of new sources on your same topic, and chances are there may be one that you hadn't found through other means.

And because you're using the bibliography of a source that had some relationship to your topic, you have a better chance of striking gold with something on that list. Even *Wikipedia*, which is questionable as a source by itself, can prove fruitful: if you check out the pages on Ebola, for example, and look for the numbers at the ends of a sentence, that indicates that the sentence has a source for its information. That source is listed at the bottom of the page in the "Reference" section:

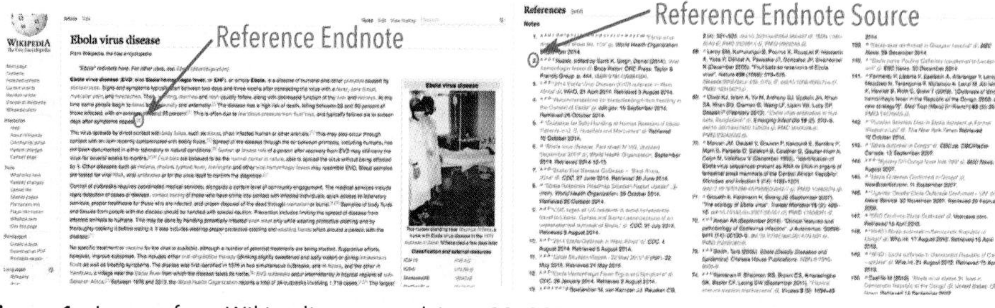

Figure 1. Images from Wikipedia, accessed June 28, 2014.

Often, that source information includes a hyperlink that allows you to link directly to the original primary or secondary source on the Wikipedia page. But you have to be careful: sometimes, the link is to a personal blog or website instead of a reputable organization or agency. Before you use that information, double-check the source itself for credibility and accuracy.

Pain of Tossing out Sources

So here's an uncomfortable fact about research that many people don't like to talk about: yes, you are going to have to read things you won't end up using in your paper. It's difficult to admit, but you will spend valuable time throughout your college career reading a host of sources that you will realize, in the end, aren't what you're looking for, and you'll have to scrap them. The first time you do this, it will feel a little bit like throwing away a $20 dollar bill (after all, your time is worth something, and you just flushed it away!).

But what you'll discover is that learning to let go sometimes will help you to become much more effective at using evidence well. The pain of tossing away a source that you've spent time on is counterbalanced by your growing understanding of how to distinguish valuable sources from ones that aren't as valuable. And, in the end, you'll become a better writer, because your professors will see that you didn't just throw in every source and the kitchen sink to get a paper done, but instead you tried to choose the best possible sources to bolster your particular claims. And getting the good grades that come along with that realization is a pretty great pain reliever.

If a source does work for you, again, the bibliography also might.

Just as the bibliography of a source that you discard might be valuable, so might the bibliography of a source that you deem helpful—likely, even more so. Be sure to read through them to see what other research gems you might uncover that you may have overlooked.

Make sure you are using a diversity of sources.

Although you may have gathered 10 sources, if they are all sources by the same author or from the same publication, you're not getting a representative sampling of the diversity of research available out there on your topic. No matter how unbiased we attempt to be, humans are fundamentally biased; although we all try to keep our biases out of our work, none of us are perfect. So if you consult only one author or one publication for your information on a topic, you run the risk of falling prey to their individual biases, which might created a blind spot in how you support your claim. If you draw from a diversity of sources, however, chances are your pool of sources will balance one another out—what one misses, another will see clearly.

Make sure your sources aren't all from the same time period.

Unless you are researching a topic whose development stalled during some period of time, it's generally not a good idea to have sources that were all written during the same period. If you do, generally you'll end up missing critical information.

For example, if you chose to not to read any research on Al Franken beyond what was published about him through 2007, you might still think he was a comedian, talk-show host, and former *Saturday Night Live* writer who had an interest in stirring the political pot with his writing and commentary. But if you expanded your research beyond that year, you'd learn that Franken ran for Senate in Minnesota and was narrowly elected to the office in 2008, where he still serves today. As you can see, it's incredibly important to broaden the pool of your research to cover both historical and contemporary time periods because a lot can change on a topic in just a few short years.

After you've noted all of these issues, looks for obvious gaps in your research.

Once you've narrowed your list of sources and noted the patterns listed above, look for the gaps and try as best you can to fill them in. If there's a gap or a concentration for some reason, make sure you can justify it—for example, if you have a host of valuable sources from the year 1996, maybe there was a surge in research on your topic that year that advanced knowledge about your topic in some fundamental way. Or if you have several sources from the same author, perhaps that person is a noted and reliable scholar who deserves a little more textual landscape than some others. Don't be afraid of gaps and don't feel like your sources have to be perfectly balanced—just be prepared to justify those gaps to yourself and your audience.

Case Study 1: Judith Miller's Reporting on Weapons of Mass Destruction

Even researchers with a reputation for vigilance can sometimes make mistakes when it comes to distinguishing reliable sources from unreliable ones. Judith Miller, a Pulitzer Prize winning reporter for the *New York Times*, found that out the hard way.

Figure 2. Judith Miller (NYT).

Miller, in conjunction with journalist James Risen, won the Pulitzer Prize for Explanatory Reporting in 2002 for their 2001 coverage of their coverage of global terrorism before and after the September 11 attacks. Despite winning the highest accolade for journalism in the country, Miller was still susceptible to making a mistake with a source. Much of her coverage of the Iraq War and the country's supposed "weapons of mass destruction" in 2002 and 2003 was later proven to have been based on incorrect information.

The source of at least some of that incorrect information was Ahmed Chalabi, an Iraqi politician and former oil minister who was initially a trusted intelligence informant for the United States government. Chalabi's information linking Saddam Hussein to Al-Quaeda and weapons of mass destruction, however, was later proven false, and his position as informant and his relationship to the United States grew sour.

Trusting the wrong source had a dramatic impact on Judith Miller and on her career: by October 2005, even the *Times* Public Editor Byron Calame was stating publicly that because of the inaccuracies in her reporting, it would be "difficult for her to return to the newsroom as a reporter."

For any other reporter, this kind of inaccuracy would probably be a career-ender, but winning the Pulitzer Prize does still hold some weight: Miller continued to write for other publications and eventually was offered a job at Fox News, which she still holds today. Still, the legacy of not being discriminating enough about her source is one that will follow her name throughout the remainder of her career and likely her life.

Read more about Judith Miller here:

From the *New York Times*, **"Threats and Responses: The Iraqis; US Says Hussein Intensifies Search for A-Bomb Parts"** *(http://www.nytimes.com/2002/09/08/world/threats-responses-iraqis-us-says-hussein-intensifies-quest-for-bomb-parts.html?module=Search&mabReward=relbias:r,[%22RI:5%22,%22RI:18%22]).*

From the *New York Times*, **"The Miller Mess: Lingering Issues among the Answers"** *(http://www.nytimes.com/2005/10/23/opinion/23publiceditor.html?pagewanted=print&_r=0).*

Exercise: Case Study

Read some of the additional reporting from and about Judith Miller's reporting on Iraq in the early 2000s. What could and/or should have Miller done differently in order to verify the accuracy of the information she was receiving from Chalabi?

Consider your own approach to verifying information for a paper. What are you doing now to make sure that the information you are gathering is accurate? What could you do differently to improve upon this process?

STEP TWO—REFLECT: WHAT DOES IT MEAN?

Now that you have a good idea of what kind and what variety of sources you have, you also need to have a good understanding of what those sources are saying. That stack of books or articles might be intimidating, but again, you'll have a much better grasp on where your paper is headed if you have an understanding of what the larger academic and cultural conversation is about your topic. So, again, before you start to write, make sure you **understand the content of your sources.** Here are a few tips that will help you move through that process smoothly and effectively:

Understand the ethos of your source's author.

Before you even begin reading a source, it's a good idea to do a little cursory reading about the author. Read their bio in the back of the book; perhaps Google them quickly to see what you can find. As we noted in Step One, we're all biased—it's unavoidable—so it might be a good idea to know what your author's history is related to the subject they are writing about.

For example, Dr. Mehmet Oz, host of the "Dr. Oz Show," might seem like a reputable source to quote on the medical value of green coffee beans or Garcinia cambogia. But a little research shows that he has been called before a congressional subcommittee on consumer safety, and the committee has stated that both green coffee beans and Garcinia cambogia have been promoted as tools for weight loss on Dr. Oz's show, but that neither product has any scientific evidence to back up those statements.

Don't quote from sources with which you aren't familiar.

It might seem easy to simply flip to a page in a source and to pull a quote out to use in your paper, but it's also a very risky maneuver. When you pull out a quote at random that looks good—an action called "contextomy"—you risk two important things. First of all, you risk misunderstanding the context of that quote. What that means, in essence, is that if you haven't read the entire chapter or section that a quote comes from, chances are you're going to miss an important part of the argument. Second, if you don't read the entire article or book, there's a good chance that you're going to miss out on a quote that might be even better than the one you're choosing.

Similarly, have a good understanding of each source's argument.

This is a challenge, but it's absolutely worth it, and it goes hand-in-hand with the previous suggestion. Make sure that you actually read your sources and that you try to pinpoint the arguments the authors are making. If it is an article, what is the one takeaway that the author is attempting to prove? If it is a book, what is the overall point that ties the chapters together?

Learn how to skim and scan effectively.

Reading an article or two for an essay seems pretty reasonable, but an entire book? Or two? Or more? This kind of research is daunting to most students to say the least. But learning to skim and to scan effectively will help you move through large volumes of text while still gathering pertinent information.

Skimming is essentially a method of reading where you move your eyes quickly over a source while attempting to gather information that will help you better understand the source's main points. Read for these informational clues while skimming:

- The source's title
- Chapter headings
- Sub-headings for sections within chapters
- The introductory paragraph in full
- The first and last sentence of each paragraph (if the source is not book-length)
- The concluding paragraph in full
- Unusual words
- Words that provide concrete answers to the "five w" questions and "how"
- Any words that are italicized, in bold, or set apart in any way
- Any graphics

Scanning is essentially a method of reading where you read over large volumes of information in a source while looking for particular information. Read for these informational clues while skimming:

- Main ideas (typically in an introductory paragraph or a topic sentence)
- Facts or statistics
- Quotes
- Dates
- Other information related to a person, thing or event

Take notes on your sources.

Reading through your sources is valuable, but when it comes time to write your paper, you'll frustrate yourself to no end if you don't have notes on those sources to which you can refer. Note-taking, which can include writing down memorable points on a separate piece of paper or adding margin notes to a text (also called annotation), is a valuable skill that not only gives you a quick reference point for what you've read, but it also helps you to retain that information a bit better. Here are a few tips on note-taking and annotation:

- Take notes on the same categories and clues that you skim and scan for.
- Make sure to write down page numbers for quick reference.
- Use different-colored highlighters to identify different points in the text.

- Write down a summarizing word or phrase next to a particularly relevant paragraph or passage.
- At the end of your notes on a source, write down a quick paragraph or sentence synopsis with the most important take-away.
- Group the notes of similar sources together.

Case Study 2: Media Misunderstands Supreme Court's Obamacare Vote

Major news networks usually take great pains to get their facts right, but sometimes even they misunderstand their source material in their haste to get a news story out to the public. The Supreme Court's June 28, 2012 decision on the Affordable Care Act, known colloquially as "Obamacare," is case-in-point.

The Supreme Court's decision on Obamacare was revolutionary in a number of ways: it upheld a very controversial piece of legislation that changed Americans' relationship to their healthcare and their healthcare system, but it also showcased a major flub on the part of two prominent media outlets.

In the minutes following the release of the decision, networks and media outlets—including Reuters, the Associated Press, and Dow Jones—attempted to wrestle their way through the nearly 200-page decision and concluded that the Supreme Court had upheld one of the legislation's most controversial components: the individual mandate that requires all United States residents to have health insurance. Both CNN and Fox News reported that the mandate had been struck down. Why did this error occur? There's no doubt that the reporters and producers on the story were rushed and eager to get their information out first and didn't take the time to review the primary source document—the ruling itself—in enough detail. According to a story issued by Poynter on June 28, CNN issued a correction about 90 minutes after the error occurred and said that the network "regrets that it didn't wait to report out the full and complete opinion regarding the mandate." Fox News also issued an apology to its viewers.

This kind of error by major media organizations illustrated that no one is immune to making this kind of mistake. CNN was embarrassed publicly and compounded its problems by not only reporting the error on the ruling on its live broadcast, but also on its website, by email, and on its social media sites. Its tweet about the ruling, for example, was retweeted hundreds of times by the time the network realized its error and sent out a correction over 10 minutes later. Fox News was also embarrassed by its error on live tv, but because at the time it hadn't fully integrated its method of distributing information digitally, it did not deal with as many retractions as CNN had to cope with. The moral of the story? Make sure you read your source thoroughly and get your information right, even if you don't get it out first.

Figure 3. Poynter's screenshot of CNN's website.

Read more about CNN and Fox News' error on Obamacare here:

From Poynter.org, **"CNN, Fox News Err in Covering Supreme Court Healthcare Ruling"** *(http://www.poynter.org/news/mediawire/179144/how-journalists-are-covering-todays-scotus-health-care-ruling/).*

From SCOTUSblog, **"We're Getting Wildly Different Assessments"** *(http://www.scotusblog.com/2012/07/were-getting-wildly-differing-assessments/).*

Exercise: Case Study

1. Read the SCOTUSblog article, particularly the section under the time 10:08:30 where journalist Tom Goldstein talks about his skimming approach to reading the Supreme Court's opinion. What did he do well in this exercise? What could he have improved upon to make his skimming more effective?
2. Consider the apology that CNN and Fox News offered their viewers after they realized their mistake. What would you do if you realized that you had made a serious error in your assessment of a source for a paper? How would you approach your professor to discuss it? And how could you keep yourself from making that kind of error in the future?

STEP THREE—RESPOND: WHAT DO I DO WITH IT?

Now that you've evaluated your sources and understand their content, you're ready to start writing. Once you get your introduction and your thesis crafted, you'll probably be looking for source material to use to bolster your first point, so you're also ready to begin to **integrate sources into your paper.** This is an exciting point in the writing process, but it's also tricky—many students get frustrated when they begin to put source material into their work, because they're not quite sure how to do it effectively. Here are a few tips for you on how to make sure that that source material works to both bolster your claim and push you to make it even stronger:

Draft freely.

Although this isn't a point that is directly related to using evidence, it's a point that, if executed well, will allow you to integrate your evidence more effectively. Try your best to turn off your internal editor and just write. Get your ideas down on paper to the best of your ability so that you aren't judging your work harshly while it is still in an early draft stage and so that you can feel confident about your initial ideas. This will give you more courage to reach out during the revision process to integrate your source material.

Use shorthand to designate places in your paper where you'd like to include source material.

Don't feel compelled to integrate your source material seamlessly on the first try. Most effective writers have worked with and revised their source material numerous times in order to integrate it smoothly. Sometimes the easiest way to begin the process is to simply put a note to yourself in the text of your paper about what source you want to use and what point you'll be highlighting from it in parentheses next to your claim. You can come back later and work on polishing it.

Let your sources inform your thesis and your claims.

When you write a paper, remember that you aren't starting off with an ironclad idea for which you need to find absolute and unequivocal support. Your thesis is not static—it's elastic, and it should be flexible and receptive to change based on what source material you've discovered. In other words, you aren't making an inflexible argument and then looking for sources to support that idea—you are actually testing your argument similar to the way that scientists test a hypothesis; and as a result, you need to be willing to adapt that argument to new discoveries and information just as scientists do. That does not mean, however, that you shouldn't have confidence in your argument or that you should scrap it at the first sign that one of your sources might disagree with you. You are looking for support, but you are also looking for people who might have more in-depth information that you do to inform your argument and improve it.

Don't let your sources overburden your work.

Excellent source material is great, but overkill is also always a possibility. A paper that relies too much on source material can cause the voice of the paper's author to disappear. This happens sometimes because people don't have enough confidence in their ideas, so they lean on source material as a kind of crutch. It's very important to place your ideas in the larger conversation about your topic, but you don't want your ideas to be completely drowned out. The answer? Make sure that you surround every instance of source material with your own ideas. We'll explore some concrete ways to do this below.

At the same time, don't lose sight of your sources.

The flip side of the problem above is that some students don't rely enough on the larger conversation on their topic and can end up with an argument or ideas that get out of their control. Work hard to maintain your focus on your ideas and how your source material can support those ideas, because if you deviate too far, you could lose control of your paper and your message.

Case Study 3: CaroMont Regional Medical Center's Rebranding Campaign

There are a number of examples out there that illustrate how easy—and damaging—it can be to lose sight of your source material, but the example of North Carolina's CaroMont Regional Medical Center's rebranding campaign is a painful and memorable one.

In the spring of 2013, CaroMont—located in Gaston, North Carolina—announced a name change for the Gaston Memorial Hospital and a rebranding. The hospital, which would now be known as CaroMont Regional Medical Center, would also abandon its former tagline "In Love With Life." Its replacement? "Cheat Death."

Figure 4. The Gaston Gazette's photograph of the infamous slogan reveal. (http://www.gastongazette.com/caromont-rethinks-cheat-death-updated-1.123327)

The new tagline drew rapid and unrelenting critique from members of the public and from national news media outlets like *The Huffington Post* and *NPR*. Some were merely uncomfortable with the tagline's negative connotations; others mocked it and treated it as if it were a joke. Local residents, however, who had lost relatives at the hospital felt like the message was both demeaning and insulting: After all, they said, no one cheats death, and nowhere is that more apparent and more inescapable than at a hospital.

So what went wrong with this new campaign? In part, the fault lay with the hospital's board of directors who gave the job of creating the new tagline to Immortology, a Chapel Hill-based marketing and branding firm which many said did not do a sufficient job of vetting the proposed rebranding internally, much less among public focus groups, before launching it.

But also problematic was the fact that the impetus for changing the tagline was lost in the shuffle. Hospital board members had recently been made aware that Gaston County had scored near the bottom of the state's 99 counties and on a national level on an assessment of community health conducted annually by the University of Wisconsin Population Health Institute and the Robert Wood Johnson Foundation. A new tagline and associated campaign, they thought, just might spur county residents to improve their health.

What made headlines here, however, was not the good intentions of the hospital board but instead the failed attempt of a firm that, according to its website, believes that "safe is for sissies," a firm that creates only "immortal" brands that "lead rather than follow." Unfortunately, the only thing that's lasting about this particular story is a public relations disaster that the hospital will never quite live down.

Read more about CaroMont's PR disaster here:

From *Charlotte Business Journal*, **"CaroMont's 'Cheat Death' Slogan Drawing a Sharp response"** *(http://www.bizjournals.com/charlotte/news/2013/04/05/caromonts-cheat-death-slogan.html?page=all).*

From *Modern Healthcare*, **"Outliers: New Slogan Dies Quick Death"** *(http://www.modern-healthcare.com/article/20130413/MAGAZINE/304139958).*

Exercise: Case Study

In the *Modern Healthcare* article, CaroMont CEO and President Randy Kelley says the hospital's intent was "never to offend or incite." How do you think CaroMont could have handled this differently that would have kept the public's focus more on the health report that spurred their desire for change?

Consider what would happen if you lost sight of source material in a paper. What steps could you take to revise your paper so that your argument connected effectively with your source material again?

For every major claim you make in support of your thesis, try to have at least one source.

This is less a hard-and-fast rule than a suggestion. If possible, try to see if you can find a source that will support each major claim that you make in your paper. This is a challenge, but if you aim for this standard, it will help you maintain balance in your paper between reputable sources and your own ideas.

Make sure you indicate whether your source material is quoted, paraphrased, or summarized.

It's very important to indicate in your paper whether you are using source material that is a quote, is being paraphrased, or being summarized. Here's a brief description and example of each to help you understand the difference between these classifications:

- **Quoted source material** is material that is replicated word for word in your paper. This material is always set apart by quotation marks.

  Example: In his 1955 essay "Stranger in the Village," James Baldwin wrote "but some of the men have accused *le sale negre*—behind my back—of stealing wood and there is already in the eyes of some of them that peculiar, intent, paranoiac malevolence which one sometimes surprises in the eyes of American white men when, out walking with their Sunday girl, they see a Negro male approach."

- **Paraphrased source material** is generally a passage whose general intent and meaning is restated succinctly and faithfully in the writer's own words. Paraphrasing is typically confined to a small amount of quoted text. This material is also usually afforded some kind of introduction, and no quotation marks are typically used unless the writer feels the need to retain some of the author's original vocabulary.

  Example: In his 1955 essay "Stranger in the Village," James Baldwin wrote that male residents of the Swiss village still accuse him of stealing behind his back and silently radiate the same fear and hatred that he has seen in the eyes of American men.

- **Summarized source material** is material that attempts to put the main ideas of a text succinctly and faithfully into the writer's own words. Typically, summarized source material incorporates a much larger sections of text, such as a paragraph, chapter, article, or book. As with others, this material is also afforded some kind of introduction, and no quotation marks are typically used unless the writer feels the need to retain some of the author's original vocabulary.

  Example: In his 1955 essay "Stranger in the Village," James Baldwin wrote about the unsettling experience of being the only African American man in a village of 600 white Swiss villagers, the odd quietude associated with living in a remote village in the Alps in the middle of winter, the dramatic difference between the village and his hometown of Harlem, and the inescapable history of race and racism.

> To integrate a quote, paraphrase, or summary effectively, follow these three steps: introduce, insert, and explain.

The process of quote integration can be a bit tricky, especially for new writers. Although there's no real formula for doing this effectively, we've outlined a three-step process below with examples that can serve as a guide. Once they are comfortable with this approach, advanced writers can feel free to deviate from this process and use these steps in varying ways to help them construct a more complex argument.

Introduce

Try offering a two-part introductory statement that indicates to your reader important information about your source material:

Part I

This part of the introduction should introduce us to the text. It can include the title, date, or author, or if it feels relevant or important, all three.

Example: In *The Great Gatsby,* written by F. Scott Fitzgerald in 1925, …

Typically, you only need to offer this much information once in your paper. After the first mention, and any time you include another quote, you can simply refer to the title of the book or the author or, in some cases, you may not need to mention the title at all because it will be understood contextually.

Example: Fitzgerald's novel … or In the novel …

Part II

This part of the introduction should introduce us to the relevant quote itself. It should be a statement that gives us the context for the quote.

Example: In *The Great Gatsby,* written by F. Scott Fitzgerald in 1925, Gatsby's great love Daisy Buchanan and her friend Mrs. Baxter are introduced to the reader …

Insert

Adding Citation

After you've written your two-part introductory statement, you are ready to insert your quote, paraphrase, or summary. Make sure you follow all correct citation rules (see Step Four—Revise: How Do I Make It Better? on pg. 178).

Example: In *The Great Gatsby,* written by F. Scott Fitzgerald in 1925, Gatsby's great love Daisy Buchanan and her friend Mrs. Baxter are introduced to the reader "in white, and their dresses were rippling and fluttering as if they had blown back in after a short flight around the house" (27).

Explain

Now that you've given relevant information about the text, relevant information about the quote, and provided your quote, you can provide a two-part explanation for the quote.

Part I

This part is your explanation of what the quote means to you. Consider the details, images, and description in the original quote and try to put them into your own words and to explain their meaning.

Example: In *The Great Gatsby,* written by F. Scott Fitzgerald in 1925, Gatsby's great love Daisy Buchanan and her friend Mrs. Baxter are introduced to the reader "in white, and their dresses were rippling and fluttering as if they had blown back in after a short flight around the house" (27). This description depicts Daisy and Mrs. Baxter as birds, swans perhaps, elegant in white, ornamental, flighty, unable to stay in one place for too long.

Part II

This part is your explanation of why this quote is relevant to your claim. How does it support your argument? Err on the side of providing too much explanation, if possible; it's always easier to cut text out of a paper rather than try to come up with additional information.

Example: In *The Great Gatsby,* written by F. Scott Fitzgerald in 1925, Gatsby's great love Daisy Buchanan and her friend Mrs. Baxter are introduced to the reader "in white, and their dresses were rippling and fluttering as if they had blown back in after a short flight around the house" (27). This description depicts Daisy and Mrs. Baxter as birds, swans perhaps, elegant in white, ornamental, flighty, unable to stay in one place for too long. Although it seems at first like merely a frivolous and perhaps derogatory depiction of the novel's women, in fact it hints at the the true nature of Daisy's character—her lack of sincerity and focus and her utter inability to care about more than appearances—that makes her one of the most damning and damaging characters in the novel.

Avoid these common missteps because they violate the three-step process for source material integration.

There are lots of rules that we could cite here, but these are three common mistakes that new writers make that run contrary to the three-step process we've just outlined. Read them through so that you can be aware of them and avoid them.

- **Try not to let a quote stand on its own in a sentence.** Some writers feel like they shouldn't tamper too much with an author's words because they speak for themselves, but that's not the purpose of using them in a paper. You are using them to bolster your argument. In order to do that effectively, you have to provide context and connections to what you're writing, so it's much more effective to provide an introduction to a quote that links it seamlessly to your work.
- **Don't end a paragraph with a quote.** It's another common mistake that new writers make: end with a quote, and you'll end with the strongest possible sentiment, right? Wrong. In the end, you end up weakening your own argument because you're relying on the words of someone else to fully explain your point rather than using those words to supplement and bolster your own. Whenever possible, try to end with some kind of explanation of the quote.
- **Don't use quotes in your topic sentences.** Topic sentences are places for you to work your writerly magic—not places for the words of others. These are the sentences where you need to set forth your claims succinctly using your own thoughts and ideas. If you use source material here, you run the risk of having that source material monopolize the conversation.

Attribute your material appropriately—or, in other words, cite your sources.

This rule is one that's paramount. Just as you want to be given credit for your original words and thoughts, so do others. Make sure that you always cite words and ideas that are other than your own. And if you're not sure you're doing it correctly, look it up.

Case Study 4: The Fabrications of Stephen Glass

Most people think of plagiarism as primarily a problem among high school and college students, but not so: in fact, there is a long and unsettling pattern of plagiarism and fabrication in the professional world, particularly in the field of journalism, that can teach us some valuable lessons about why it's so important to cite the work of others—and to tell the truth.

One of the most infamous examples of this rising trend is *Republic* magazine. One of Glass' most appealing qualities was his uncanny ability to find the most incredible stories: about teenage computer hackers who extorted money from corporations and a conference for drunk and angry young conservatives that boasted orgies and drug fests. The only problem was the stories were almost all fabricated.

In fact, once the *New Republic's* editorial staff finally made their way through an investigation and fact-checking of all of Glass' 41 bylines for the magazine, they discovered that 27 of his pieces contained fabrications. The staff noted that in a few cases, entire stories were invented.

Figure 5. Stephen Glass photo used in 1998 *Vanity Fair* profile. Photo by F/Stop Studios.

Glass, of course, is not the only writer to be branded as a fabulist or a plagiarist. Former *New York Times* reporter Jayson Blair and columnist Maureen Dowd have both been accused of plagiarism, and *A Million Little Pieces* author James Frey has also been accused of being a fabulist. Although there will always be people who will complete this disturbing pattern, the resulting public humiliation and guilt should be enough of a motivator to remind most of us of how important it is to credit the ideas of others and to stick to the truth.

Read more about Stephen Glass:

From *Vanity Fair*, **"Shattered Glass"** *http://www.vanityfair.com/magazine/1998/09/bissinger 199809.*

From Media Research Center, **"The New Republic Plays the Victim"** *(http://www.mrc.org/bozells-column/new-republic-plays-victim).*

Exercise: Case Study

Read "Shattered Glass." What do you think Stephen Glass' reasons were for fabricating his stories? Why not just find genuine stories to write about?

Think carefully about the pitfalls associated with plagiarism. What steps can you take when writing your next paper to keep yourself from even accidentally plagiarizing?

Know which style of documentation you need to utilize.

Most professors will tell you clearly what they're looking for, so pay attention, and if you're not sure, ask. Most humanities classes, for example, use MLA style, and most social sciences use APA. The rules for these methods of documentation are clearly outlined in numerous books and on numerous websites, so the answers you might be looking for are readily available.

STEP FOUR—REVISE: HOW DO I MAKE IT BETTER?

Your paper is drafted, so it's ready to turn in, right? Not so fast—**revising and editing** are both fruitful and necessary, and you will be well served to follow at least a few steps before turning your final draft in to your professor. In fact, many students find that their best writing happens during the revision process. Give yourself enough time to go through your paper at least once before you turn it in—you'll be glad you did.

Be your own fact-checker.

This is a simple step but incredibly important. Check for:

- Spelling, grammar, and punctuation errors.
- The spelling of names in particular.
- The accuracy of source page numbers.
- The accuracy of quotes.

Make sure your use of source material is balanced.

You don't want to turn in a paper that's heavily dominated with sources in one area and then devoid of sources in other areas. You want those sources to be used effectively and to be distributed relatively evenly so that your claims are supported equally by material. Remember—the use of sources is less about the number than it is about how effectively and artfully they are incorporated.

Polish your integration of sources.

In other words, make sure that when you integrate your source material using accurate and effective punctuation:

- Use ellipses to shorten a quote.
- Use parentheses for in-text citations and to indicate an error in an original quote (for example, "sic" to indicated incorrect spelling).
- Use brackets to help clarify or explain a quote.
- Alter your grammar so that it matches the grammar in your paper.

Research

Chapter 9

When I first moved to Kentucky, I found myself driving to Versailles in order to visit a restaurant I had heard about. The drive was lovely and relaxing, but there wasn't much to see along the way. Suddenly, in the middle of nowhere was a giant castle that looked like it belonged in a European countryside.

Figure 1. What is this castle doing in the middle of Kentucky?[1] Image by Christina Ramey. Licensed under a Creative Commons Attribution-Share Alike 2.0 Generic license. *https://flic.kr/p/5BBNGe.*

"Did you see *that?*" I asked my husband.

"I have no idea," he said. "What the heck *is* that thing?"

When we arrived at the restaurant, I immediately pulled out my iPhone and did a quick online search to try and figure out what a giant castle was doing in the middle of Kentucky. I chose some keywords that I thought might get me the results I wanted.

I typed in "*castle + Versailles + Kentucky.*" The first several links that came up were links to sites about something called "CastlePost" and a *Wikipedia* entry for "Martin Castle."

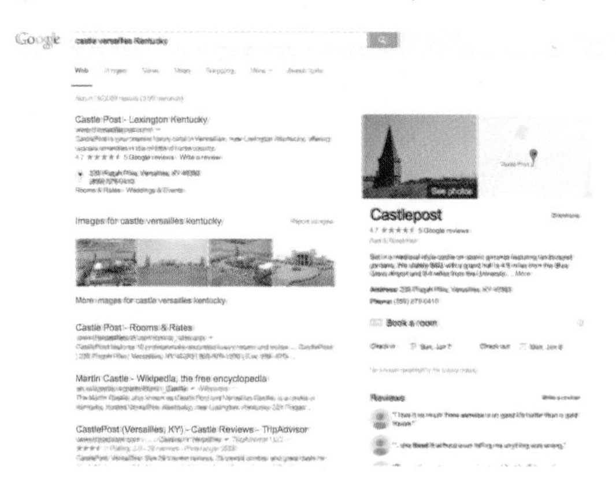

Figure 2. What your Google search might look like if you searched for information about Kentucky's mystery castle.[2]

I clicked on the Wikipedia entry and discovered that this was the strange site I'd seen. According to the Martin Castle Wikipedia page, the castle was built in 1969 in Versailles, Kentucky. A local contractor named Rex Martin built the castle after returning from a trip to Europe, where he was smitten with the castles dotting the European landscape. Unfortunately, Rex Martin and his wife divorced in the 1970s and the castle was left unfinished. It still sits, relatively unused, on the roadside in Versailles, Kentucky.[3]

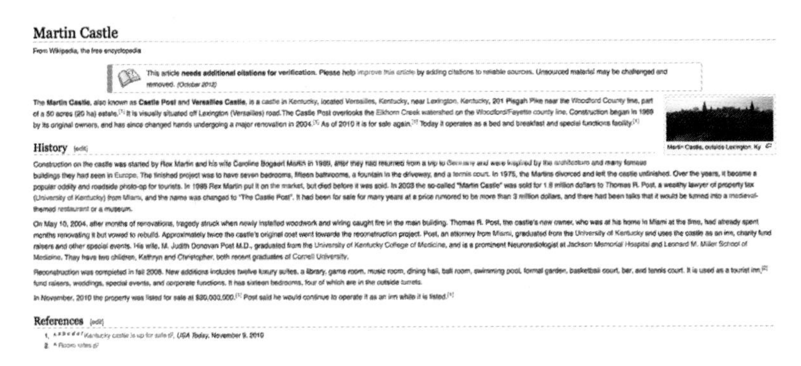

Martin Castle

From Wikipedia, the free encyclopedia

This article **needs additional citations for verification.** Please help improve this article by adding citations to reliable sources. Unsourced material may be challenged and removed. *(October 2012)*

The **Martin Castle**, also known as **Castle Post** and **Versailles Castle**, is a castle in Kentucky, located Versailles, Kentucky, near Lexington, Kentucky, 201 Pisgah Pike near the Woodford County line, part of a 50 acres (20 ha) estate.[1] It is visually situated off Lexington (Versailles) road. The Castle Post overlooks the Elkhorn Creek watershed on the Woodford/Fayette county line. Construction began in 1969 by its original owners, and has since changed hands undergoing a major renovation in 2004.[1] As of 2010 it is for sale again.[1] Today it operates as a bed and breakfast and special functions facility.[1]

Martin Castle, outside Lexington, Ky

History [edit]

Construction on the castle was started by Rex Martin and his wife Caroline Bogaert Martin in 1969, after they had returned from a trip to Germany and were inspired by the architecture and many famous buildings they had seen in Europe. The finished project was to have seven bedrooms, fifteen bathrooms, a fountain in the driveway, and a tennis court. In 1975, the Martins divorced and left the castle unfinished. Over the years, it became a popular oddity and roadside photo-op for tourists. In 1988 Rex Martin put it on the market, but died before it was sold. In 2003 the so-called "Martin Castle" was sold for 1.8 million dollars to Thomas R. Post, a wealthy lawyer of property tax (University of Kentucky) from Miami, and the name was changed to "The Castle Post". It had been for sale for many years at a price rumored to be more than 3 million dollars, and there had been talks that it would be turned into a medieval-themed restaurant or a museum.

On May 10, 2004, after months of renovations, tragedy struck when newly installed woodwork and wiring caught fire in the main building. Thomas R. Post, the castle's new owner, who was at his home in Miami at the time, had already spent months renovating it but vowed to rebuild. Approximately twice the castle's original cost went towards the reconstruction project. Post, an attorney from Miami, graduated from the University of Kentucky and uses the castle as an inn, charity fund raisers and other special events. His wife, M. Judith Donovan Post M.D., graduated from the University of Kentucky College of Medicine, and is a prominent Neuroradiologist at Jackson Memorial Hospital and Leonard M. Miller School of Medicine. They have two children, Kathryn and Christopher, both recent graduates of Cornell University.

Reconstruction was completed in fall 2008. New additions includes twelve luxury suites, a library, game room, music room, dining hall, ball room, swimming pool, formal garden, basketball court, bar, and tennis court. It is used as a tourist inn,[2] fund raisers, weddings, special events, and corporate functions. It has sixteen bedrooms, four of which are in the outside turrets.

In November, 2010 the property was listed for sale at $30,000,000.[1] Post said he would continue to operate it as an inn while it is listed.[1]

References [edit]

1. ^ a b c d e f Kentucky castle is up for sale, *USA Today*, November 9, 2010
2. ^ Room rates

Figure 3. What you might see if you search Wikipedia for "Martin Castle."[4]

The entry mentioned that the current owner has been criticized by locals for failing to devote much attention to the landscaping on the sprawling property. My husband and I found ourselves continuing to talk about the castle and the history we had just learned.

I asked, "What do you think people around here think about the castle?"

"I don't know," he said. "Look it up."

But how could I look up public opinion easily? My previous search didn't seem to turn up lots of details about local opinions on this huge piece of architecture. Maybe the local newspaper has a story on the castle. I knew that local papers not only have archives of stories online, but they often archive reader responses to online stories.

I went to the *Lexington HeraldLeader's* website at Kentucky.com and I immediately found a story about Woodford county approving CastlePost to open as a restaurant. There were fourteen reader comments to the story.[5]

Figure 4. Some of the comments posted on the news story about CastlePost's approval as a restaurant.[6]

Some people thought the restaurant was a good idea, but other comments showed that people in the area weren't too happy about the castle at all. One reader even called the castle a "land scar" instead of a "landmark."

As my husband and I finished our sandwiches, we talked about how funny it was that we had learned such an interesting—and strange—piece of local history just by doing a little research during lunch.

HOW DO YOU START RESEARCH?

Although my castle search might not seem like the kind of research you do in school, it really is very similar. Research isn't always complicated or hard. In fact, we do research every day without really knowing that we're doing it! Whenever I see or hear something strange, I tend to immediately reach for a computer or book so that I can look up more information about it. Think of the ways in which you already act as a researcher:

A researcher starts with some kind of question or observation:

- *That's strange! What is that?*
- *Why isn't my car starting?*
- *How do I make peanut butter fudge?*
- *Why did World War I start?*
- *Have there ever been any unmarried U.S. Presidents?*
- *Is organic food really healthier for you?*

A researcher then goes searching for some kind of answer or clue:

- Doing online searches
- Asking experts or people who might know an answer
- Reading books
- Doing experiments and trying out various possibilities
- Observing and noticing other clues

Think of how many different questions you've researched in the past week. Did you do a search for any recipes or for price comparisons on something you want to buy? Have you asked any friends about their background knowledge of a topic? Have you called your parents to ask them any questions? All of this is *research!*

Exercise: Strange Happenings

Think of the last *strange* thing you saw or heard. Would you like to know more about it (like how it started, who made it, etc.)? Do a quick online search to get the background story. You can use my castle search as an example. When you finish, share what you discovered with the class.

Research is the act of looking for information, stories, facts, evidence, histories, or details that help to answer a question or solve some kind of problem.

DIFFERENT KINDS OF RESEARCH: PRIMARY VERSUS SECONDARY

Earlier, we defined research as the act of looking for information, stories, facts, evidence, histories, or details that help to answer a question or solve some kind of problem. This definition shouldn't sound too complex, since research is something we all do every day. Broadly speaking, there are two different kinds of research: **primary research** and **secondary research.**

Primary research is information that you gather from *firsthand accounts.* You take words, quotes, observations, histories, or facts directly from the actors involved in the situation. Primary research includes interviews, ethnographies, observations, archival documents, oral histories, surveys.

Secondary research is information that has been gathered firsthand by someone else and then reported in an essay, article, book, or other publication. Secondary research includes newspaper articles, academic journal articles or essays, reports of studies or surveys, media reports, etc.

In academic situations, both primary and secondary forms of research are important. You will find that you need to do both at different times.

Exercise: Song Lyrics

Sometimes song lyrics are fun to research. Do a quick search and listen to the following songs. Each of these songs tells a story about events that really happened, even though you might not know about them. After listening to the song online and/or reading the lyrics, do some online searches with a partner and try to uncover the background stories.

- "Hurricane"
- "Ohio"
- "Sunday Bloody Sunday"
- "Biko"
- "Strange Fruit"

Doing Primary Research

Authors use primary research when their question or problem would most benefit from direct evidence gathered from **primary sources.** Authors might visit archives or somehow gather direct quotes from people who are involved in the situation under discussion. They might also survey a small number of people who are part of the population being studied. When you seek firsthand accounts about your subject, you might consider the following primary research strategies.

Sometimes primary research involves what is called "fieldwork." **Fieldwork** may sound like a very technical or complicated activity, but it's actually quite enjoyable. In fact, you probably do some kind of fieldwork on a regular basis, though you might not think of it in these terms. Any time you take a picture of something strange, cool, weird, or amazing, you are doing a kind of fieldwork.

Some of us like to share these photos online with friends. When we share these photos, we're basically saying, *"Hey. Here's something cool I noticed."* Your friends might ask questions about it or they might comment on it. This conversation might lead you to look up more information or learn something new.

Here are two photos I shared online with friends. The first photo is a hand-painted sign I noticed stapled to a telephone pole in downtown Lexington. On a piece of notebook paper were the red painted words, "TEAR THE WHOLE THING DOWN." What could this mean? Tear *what* down? Who wrote this and why? I shared the image on my Facebook page, and some friends replied that they had seen similar messages around Lexington. Together, we began to speculate on where the messages were coming from and whether or not they were a joke. I decided to keep my eye out for similar messages downtown.

Another photo I shared online was much less mysterious, but still really interesting to me. As I walked across campus one day, I suddenly noticed blue doors sitting in the middle of the sidewalk. I shared them online and many friends described their experience of discovering the doors.

The doors remained on campus for about a week, and one day I decided to sit and watch how people interacted with the doors. How many people would try to walk through them? How many people would take pictures of them? How many people would walk right past them without even giving them a glance? I sat outside with my lunch and watched the doors for about an hour. During that time, a few people took pictures and stopped to look at them. But many more people walked right by them without stopping.

I may have been killing some time during my lunch break, but I was also doing fieldwork. I started with an observation *("Hey! Look at this door in the middle of the sidewalk!"),* and then I formed a question from this observation *("Does anyone else find this door interesting enough to stop and look?").* I did a little bit of observation in order to help answer this question. I was doing research.

When doing fieldwork, you can use **observations, surveys,** or **interviews** to help answer your question. But the first step is usually noticing something interesting enough to grab your attention.

Exercise: Observations

Go for a thirty-minute stroll outside. If you have a phone or device that can take digital pictures, take that with you. Take images of anything that you find strange, curious, funny, weird, cool, interesting, fascinating, etc. Don't stop to comment on them right now. Just collect as many images (or written descriptions) of interesting things that you can find.

When you get back from your stroll, pick one or two of your favorite. For ten minutes, write as many questions as you can think of about these images. What do you want to know about them? What do they make you curious about? (Don't edit anything just yet. Write *everything* down for now.) After you finish, pick out the question that is most interesting to you. Would this make a good research question?

Using Observations

Observation can be a powerful form of research. Sociologists and anthropologists have long used observations as a central part of their studies. Sometimes the researcher will interact with the community or place being studied. For example, a researcher studying first-year high school teachers might sit in on a class being taught by a first-year high school teacher. Observations can also happen online, as well. When I wanted to study how people argue about current events online, I observed several Facebook and online groups to see how participants argued back and forth.

Before observing a specific group, however, be sure that you have permission from the individuals you are studying. *This is essential if you are participating in a group or event that involves people sharing personal information of any kind.* For example, if you want to observe a support group for teenage mothers, you would need to ask for permission in advance. Make clear why you are asking to observe, what your project includes, and how the research will be used. *(See below for model permission forms.)*

If you observe a public space (like I did when I was sitting on campus, watching people walk past the blue doors), then you do not need to gain permission. For example, let's say you are interested in observing whether people at the mall tend to text on their phones even when they are physically with other people. You might sit at the mall for an hour or two in order to observe people's behaviors with their cell phones. This would not call for permission from those individuals.

Creating Surveys

You might also find **surveys** are a good way to help you discover more information about your topic. Before you decide to do a survey, you should ask yourself the following questions:

- Is there any other way of finding credible sources who might have done similar research?
- Is the audience I'm surveying likely to give me a unique perspective or insight into this topic?
- Am I asking the right people?

If I am interested in finding out whether or not Kentucky's horse racing industry is growing or shrinking, I probably wouldn't find it very helpful to survey a group of 100 random UK students. On the other hand, I could survey a group of people who participate in the horse racing industry to get a unique perspective on the topic.

You should also consider whether or not you are asking questions about **opinions,** questions about **behaviors,** or questions about **demographic information.**

- Questions about **opinion** ask about people's perspectives and opinions on certain topics. This might help you to better understand what a group thinks about your topic, but remember that the responses might not be factually correct. For example, if you survey 100 UK students to ask whether they think a Caesar salad has more or less calories than a hamburger, 80% might say that they believe that a Caesar salad has fewer calories than a hamburger. But (depending on the salad), this is not always the case! Caesar salads can easily have more calories than a Big Mac®!
- Questions about **behavior** ask about daily routines and habits. If you want to know how many hours of sleep UK students get per night, you can give out a survey asking people to tell you their sleep average.
- Questions about **demographic information** asks background details about individuals: age, gender, number of siblings, birthplace, etc.

In order to create strong survey questions, keep in mind the differences between **open-ended** and **closed-ended questions**. **Open-ended questions** are worded so that respondents actually have to *explain* their answers. These kinds of questions are best on surveys where you want the responses to offer you a narrative. **Closed-ended questions** are worded so that respondents can simply offer you a "yes" or a "no," or one-word reply.

Conducting Interviews

You can interview people directly involved with your topic, or you can interview people who may have some kind of important knowledge. Interviews are great ways of finding information from firsthand accounts. In order to have the most useful and productive interview experience, ask yourself the following questions:

Does My Topic Really Call for an Interview?

Not all topics are likely to benefit from an interview. Will interviewing a particular person give you information that you could not otherwise find? For example, if I want to know the history of the University of Kentucky, I probably do not really need to interview one of my professors. It is more likely that I can find relevant information in University-related sources, including in the archives and in publications. On the other hand, if I want to know what it is like to be a filmmaker living in Kentucky, I could probably learn a lot of talking with someone who makes films in Lexington.

How Can I Find the Right Expert?

You are surrounded by knowledgeable experts on your college campus. But how do you know the best person to interview? You will want to do some research in advance to see if a particular person has the background or expertise to answer your questions. Have they written about this topic? Do they work in a field that is relevant to your topic? Do they have firsthand experience or knowledge about your topic?

How Do I Ask for an Interview?

Let's say you have identified the perfect expert for interviewing purposes. Now comes the tricky part: asking for an interview. Keep in mind that the interviewee is doing you a favor by taking time to sit for an interview. You will need to approach the interview request with a very coherent phone call, email, or visit. Make sure you clearly explain who you are, why you are contacting them, and what you are requesting. You will also need to explain why you think they would be particularly ideal for this interview. State your available days and times, and then ask if you could arrange a time (according to their own schedule) to meet. See the example email below:

> *Dear Dr. Sykes:*
>
> *My name is Joel Tyler and I am a student at the University of Kentucky. I am currently researching the subject of amateur beekeeping in Lexington, and I was given your name as someone who is interested in beekeeping. I know your time is valuable, but I wonder if you would allow me to interview you about your hobby. Over the next two weeks, I am available every afternoon after 3:00. The interview should only take approximately 20–30 minutes, and I would be willing to meet you wherever is most convenient. If you have any questions, or if you would be willing to be interviewed, please contact me at joel.tyler27@uky.edu.*
>
> *I appreciate your time,*
>
> *Joel*

How Do I Interview Well?

A good interview starts with the interviewer's questions. Some of the worst interviews are ones where the interviewee answers closed-ended questions: "Yes." "No." "Maybe." A little boring, right? When interview questions are too closed-ended, the person being interviewed does not have a chance to elaborate or expand on her stories or ideas. By making the questions more open-ended, you invite the interviewee to share interesting, vivid, and memorable details with you.

Consider the following two interviews I conducted with former small business owners who lost their businesses. What are the differences between the two kinds of questions?

CONVERSATION 1

Jenny: *I wanted to talk with you today about your experience with owning a small business. How long did you own your business?*

Amy: *Two years.*

Jenny: *And did you go out of business for financial reasons?*

Amy: *Yes.*

Jenny: *Was it hard on your family when you had to close?*

Amy: *Yes. Very hard.*

Jenny: *How long ago was that?*

Amy: *About five years ago.*

Jenny: *Do you think you'll ever go into business again?*

Amy: *Hmm. I'm really not sure. I don't know.*

CONVERSATION 2

Jenny: *I wanted to talk with you today about your experience with owning a small business. How long did you own your business?*

Rick: *Almost five years.*

Jenny: *Can you describe the events that led to your business closing?*

Rick: *We had been struggling to pay our bills for about a year. My wife and I had to either take out additional loans, or just close the business. We stayed open for as long as we could, and then we just finally decided to give in.*

Jenny: *Do you remember the day you actually closed the business? What happened?*

Rick: *Oh sure. It wasn't really all that dramatic. We printed up a sign to hang on the door that said we were closed. We still had to go back and pack up, clean. It was pretty unexciting, except that it felt like we had just blown it.*

Jenny: *How did that struggle and the closing affect your family life?*

Rick: *There was definitely a lot of tension at home. We argued a lot because we were both stressed out, and I think we both sort of went through a mild depression when the doors closed.*

Jenny: *Can you think of one specific example of how you and your wife handled that stress differently or similarly?*

Rick: *Well, I found myself really unable to go out with friends. It was harder and harder for me to talk with people because the stress was overwhelming. I think my wife found herself wanting to talk about her stress with me, but I just didn't want to.*

My interview with Rick provided much more details and insight into his story. Amy's story, however, remains a little unclear. I did not give her an opportunity to expand on her answers. When I interviewed Rick, I tried to keep in mind some good rules of interviewing:

- **Ask open-ended questions that go beyond one-word answers.**
- **Frame your questions by asking for an example.** ("Can you think of one specific example of how you and your wife handled that stress differently or similarly?")
- **Try to ask people to describe memories of events or situations.** This helps people provide you with interesting stories and narratives that are filled with useful information. ("Do you remember the day you actually closed the business? What happened?")
- **Follow-up on responses by asking for examples.** When you get one-word responses, try asking your interviewee for a specific example of what they're describing.

Using Archives

Another good way to find primary research is to visit the archives. If you've never been to library archives, this idea might sound a little intimidating. However, archives (sometimes called *special collections* in the library) are extremely fascinating. Universities and historical societies keep historical documents and artifacts in their archives as a way to preserve the historical record about various people, places, events, and topics. **Archives** keep things like letters, original photographs, rare books, diaries, clothing, memorabilia, business records, and all sorts of other unique materials that help give researchers some insight into specific topics.

The archives at the University of Kentucky have collections of memorabilia and documents relating to a wide variety of topics, including the histories of coal mining, UK athletics, segregation in Kentucky, the bourbon industry, tobacco and hemp growing, Kentucky politics, and many, many other topics.

Before you physically step foot into any archive, it is a good idea to first visit the archive's website. Many special collections are available for viewing online, which might help to give you some additional ideas about what you want to research. For example, the **Kentucky Digital Library** *(http://kdl.kyvl.org)* is a searchable website full of historic newspapers, images, books, and oral histories.

Let's say that you were interested in researching the history of segregation in Kentucky. If you go to the Kentucky Digital Library site and type in the term "segregation," you will find over 1,200 results that include historical photos of segregated hospitals, oral histories of African-Americans who lived in Kentucky during the era of segregation, UK yearbook pages that address the issue of racial segregation, newspaper articles, and many other fascinating sources that give some perspective on the history of segregation in Kentucky. Of course, you don't need to stop there! You may contact special collections librarians directly in order to ask for help finding materials that relate to your specific topic.

Exercise: Using the Archives

Go to the **Kentucky Digital Library** *(http://kdl.kyvl.org)* right now and type "Buell Armory." What results do you see? Look to the left and identify what types of materials are available for viewing: photographs, postcards, and what else? Notice the different publication dates of the available materials. Click to view the oldest available image. What do you notice about the UK campus? What is interesting to you about this image? Click on the photo from 1918. Read the description and describe what is happening in this image.

Most students enjoy doing archival research because it's a bit like opening up a time capsule. It can be fun to find old photographs, read personal diaries from centuries ago, or even see advertisements from bygone eras. More than just a fun experience, archival discoveries can also tell us something about our history.

Take a look at these two historical advertisements, for example. What do you find interesting about them? What kinds of questions do they leave you with?

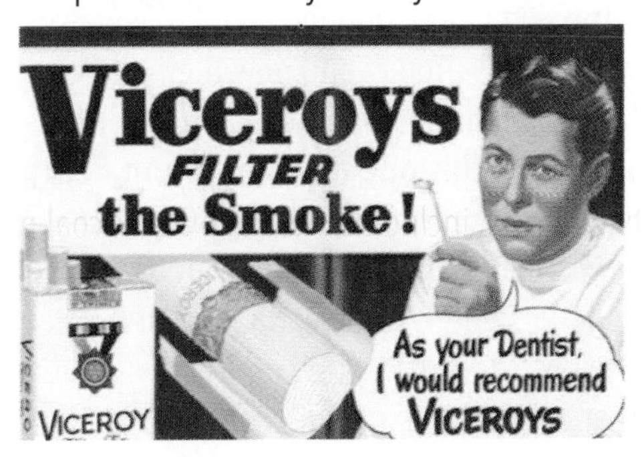

Figure 5. What kinds of questions does this historical ad for cigarettes leave you with?[8]

Figure 6. What is interesting/different about this historical ad for Wrigley's?[9]

A Word about Ethical Primary Research

Doing primary research—interviews, observations, surveys, fieldwork, archival research—is lots of fun. When you are working with other people, however, be sure to approach them in an ethical and respectful manner. Be sure to recognize that your interviewees are sharing something important with you.

- Always ask for permission before quoting an interviewee in your text. Make sure that they know who will see this project and how their names or identifying information will be used. Before beginning to work with any interview subjects, please ask them to a permission form, **like this one from the Nunn Center for Oral History at UK** *(http://libraries.uky.edu/forms/NunnCenter-Release-Master-2012-01.pdf).*
- If you are dealing with very sensitive topics or groups that may need an extra level of protection (e.g., children, elderly people, trauma survivors, etc.), please talk with your instructor about the best ways of working ethically with them.
- Always be honest and respectful in your observations. Remember that you probably wouldn't like someone to stand over your shoulder and suddenly start making notes on what you're doing. Conduct your research in a way that does not make anyone feel like a "lab rat."

Student Writing Example: "The Men Behind the Dust" by Sheldon Parrott

University of Kentucky undergraduate student Sheldon Parrott wrote "The Men Behind the Dust" for his WRD course. This ethnographic essay, which examines the lives of coal miners in Harlan County, Kentucky, draws upon primary research methods, including interviews, pictures, and observation. As you read Parrott's essay, identify the different kinds of research he uses. How does the primary research enhance your overall experience as a reader? Do the interviews and pictures give you a richer sense of the details?

Doing Secondary Research

Authors use **secondary research** when their question or problem would benefit from the findings and conclusions gathered by experts, journalists, or others who have collected and analyzed research on the topic. Authors find credible sources that summarize and describe the primary research that someone else has done. Some of the most common forms of secondary research include:

- **Academic journal articles:** Academic journals are places where academic scholars share the results of their studies and their research. These are not "popular" journals, but are written for an audience of fellow academics and researchers interested in that scholarly field. Academic journals can be helpful to you because they often share results of important research. However, because academic journal articles are written for professional academics who are already in those fields, these articles are usually more technical and detailed than a popular magazine article.

  You can find academic journals in particular fields through the library's online databases. Several of the most popular databases are easy to find from the main UK Library homepage. Each of these databases can be used to search for a huge number of journal essays. Each database hosts slightly different journals, so you may want to search in a few of them to find what interests you:

  - Academic Search Premier
  - EBSCOhost
  - JSTOR
  - LexisNexis Academic

- **Newspaper and magazine articles:** Newspapers and magazine articles can also be useful sources of information for a number of reasons. Not only do newspapers and magazines report current events, but many of them offer argumentative journalistic pieces that offer perspective on various issues. For example, the editorial pages of the *New York Times* have featured essays written by famous politicians, celebrities, and even past U.S. Presidents.
- **Books and other media:** Certain topics may lead you to find books and other media (like documentaries) that address some of your questions. Be sure to pay close attention to the author and the time of publication. If the book (or documentary) is not new, have there been later developments on this topic that you should consider?

Exercise: Using Primary and Secondary Research

Together with a partner or group, come up with one question or problem you would like to explore. This can be a big issue (like whether or not the University of Kentucky can afford to lower tuition for students) or a small issue (like what the best cheap restaurant is close to campus). Come up with two primary sources and two secondary sources that might help you to answer this question.

Which of these sources might help to answer your question most effectively?

Sites for Secondary Research

Libraries

Like all research universities, the University of Kentucky contains a number of libraries, archives, and collections, both general and particular to disciplines and professions. Information about them is readily available: on the Internet, you can follow a link to Libraries from the UK home page (head to the Site Index, look under "L"), or go straight to **the Libraries homepage** *(http://libraries.uky.edu),* and surf to the peripheries of collections here and elsewhere.

We needn't duplicate information that's so readily accessible elsewhere. Rather, we'll take a verbal tour of libraries most likely to be of use to you in your writing courses, giving impressions of their character and possibilities—since the very act of embarking on secondary research, of wading into the stacks, is a powerful form of direct experience, if you keep your eyes open.

The big one, clearly, is the **William T. Young Library.** It looms up in its own broad lawn like a giant horse stable, dwarfing the adjoining pasture. If you're new to UK, you're advised to wander in there right away. You'll see that the stable motif is no accident: consider the tapestry mural of the philanthropist Young stroking his horse's nose, located to the right of the long, curving circulation desk. Then look up toward the faraway ceiling, the glassy cupola above, cathedral-like. If the configuration of a stable were elevated to the condition of a shrine, in some religion indigenous to the Bluegrass state, this library could serve as its St. Peter's.

What's enshrined here is not thoroughbred, but rather the thoroughness of scholarship, which seeks to preserve everything. With the advent of electronic archiving devices—first microfilm, now digital—that take up next to no space, this impulse grows less physically obvious. But in a big library like this one, it's still imposing. Take a stroll through the stacks of periodicals, which fill the third floor. This is *time travel*, the closest you'll get. Not only do you have the records, the information and reflections of decades before you, you have the very objects these past people handled. Find an old magazine of news, entertainment, popular science, design, something from when your grandparents were young or before; pull out a bound volume, sit on the floor, and leaf through it. These words and images were on coffee tables, in train stations, before other eyes, on vanished tongues. It's another world—the roots of yours.

You can rest assured that the moving stacks on their little trolley tracks will not crush you as you browse: they are fitted with electronic eyes and will discern the intrusion of your form. Once you figure out what you're after, you'll press the lit arrow buttons and shift those stacks with purpose.

If you want to take notes and copy passages in the stacks, a laptop computer may come in handy. You can check one out at AV Services, on the basement level. Show a valid UK photo ID and the machine is yours for four hours, renewable if nobody's waiting for one (no reservations; first come, first served). Return it on time, as the late fee is steep, more than ten dollars an hour.

While you're in the Young Library, head down to the basement and locate **the Hub,** which houses the **Writing Center.** It's not easy to find, tucked in one of the classrooms on the side (**B-108B**), but there are signs to direct you, and once you reach it, it's a pleasant space, with friendly, accomplished writing consultants awaiting you there, smiling and with red pencils poised. It's worth seeking out.

Though UK's more specialized libraries (at the Law School, the Medical School, the College of Agriculture, the Equine Research Center) aren't the sort most undergraduates will frequent, a few warrant special mention, since their holdings may pertain to research your writing course requires or invites, and they are interesting places to visit in themselves:

- The **Fine Arts Library,** closer than the Young to the campus center, is a handsome, peaceful two-level facility holding art and photography books and collections, with music and movies as well. It's a good place to study since it's not crowded, and when you're tired of reading, you can look at the pictures.
- The **Design Library** in Pence Hall houses collections on architecture, interior design, and historic preservation. As befits its subject, it's a visually arresting space: Art Deco in style with wood and wrought iron features. Since some writing instructors favor assignments that call for exploring constructed environments, the Design Library may hold information that will come in handy.

- The **Margaret I. King Library** was UK's main library until the Young Library opened. In the large room at its east entrance, across from the Fine Arts Library, there are interesting, mildly peculiar exhibits pertaining to UK and Kentucky history—worth a look with notebook in hand. At a desk you'll see the gatekeeper to the Special Collections, UK's repository for rare, antiquated, and collection-specific holdings, a place you can't just walk into but must have good reason to access. Though most undergraduates won't have such reason, you never can tell; your interests may develop in directions that lead you to the Special Collections, and you have as much right to these holdings as anyone.

Since written texts are not all that researchers look into or libraries hold, you should know about the **UK Audio-Visual Archives,** located in the William T. Young Library, basement level. It's described on the Library website as "one of the largest university-based collections of archival film, video, and audio recordings in the United States,"[10] containing local TV and radio programming, Kentucky-related independent films, and 400,000 still photos, among other holdings.

For visual resources, you can also visit the **UK Art Museum,** our primary site for art and photography archives. The Art Museum, of course, has a sizable exhibit space in the Singletary Center, with special and continuing exhibits; plus it has significant collections of archival materials that can be surveyed online and examined by request. **Its website** *(http://finearts.uky.edu/art-museum)* can brief you on these materials: Click on Collections. A few digital images are available there, too—images of major works from the gallery, plus photos from the Works Projects Administration.

Electronic Archives

Electronic resources for research are proliferating—a moving target. Our list is suggestive, not exhaustive. We'll start with archives linked to UK and Kentucky, then fan out.

Many instructors give assignments that ask students to focus on home places or sites in local communities. A rich electronic archive for such work is the **Kentucky Digital Library,** which UK manages on behalf of libraries, archives, museums, and historical societies across the state. You can reach its website at *kdl.kyvl.org,* or follow a link from the **UK Libraries home page** *(http://libraries.uky.edu/).* At this site, you'll find links to materials under eight categories: books, newspapers, journals, manuscripts, maps, oral histories, images, and finding aids. Most are comprised of old documents that have been scanned and reproduced online: articles, narratives, letters, diaries, scrapbooks, and the like, plus whole books and newspapers.

Oral histories are interviews with people conducted to preserve their memories and experience of the past, then transcribed and put online—a powerful, accessible form of primary research that students can readily use and also contribute to through their own interviews of family and community members (more on this below).

Images are photos, illustrations, and such, scanned and posted. These materials are organized and described so they can be located through

Finding aids are designed to be keyed to information of various sorts, from overviews of whole collections, to biographical information on collectors, down through descriptions of individual items. The holdings are extensive and growing—more than a thousand oral histories, 80,000 images, and hundreds of thousands of book and newspaper pages—yet easy to maneuver through and access, with images from searches delivered ten thumbnails at a time. They're fascinating to use, like rummaging through boxes in an attic.

The Kentucky collection is the tip of a growing iceberg of digital resources available through UK. Here's the simplest way to get to troves of this material:

1. Go the UK website's Site Index.
2. Click on the letter "A."
3. Scroll down to "Archives," and click on it.

From the Archives site, you can go in different directions. Of the four choices before you, two are likely of most interest.

The topmost link is the **UK Audio-Visual Archives,** the web extension of the Young Library archive mentioned above. Click on that link, then on "Collections" at the top of the screen, and *voilà!* An enormous list of collections appears, most with online images to peruse. A variety of these can be traced out to nearly any locale and period in the Commonwealth. It's worth half-an-hour of your time just to browse about and get a sense of what's there.

Bottommost of the hyperlink choices is "University Archives and Records Program." Click on this, then on **University Archives,** then on "Archival Collections/Holdings." What you'll come upon are voluminous digital materials regarding UK itself: its history, buildings, past presidents and faculty and such. Again, the best way to find out what's there is to explore the site.

The Nollau F Series Photographs are there, for instance: a remarkable batch of digitized photos showing buildings, people, features, and artifacts from UK, Lexington, and elsewhere, from the first half of the twentieth century—the work of an engineering professor who doubled as UK's first quasi-official photographer.

The Archives website links to the **UK Libraries Oral History Project,** too, through a link to the Charles T. Wethington, Jr. Alumni/Faculty Oral History Collection, named after Lee Todd's predecessor as UK President. There's a link to the Kentucky Digital Library there, and much, much more (check out the photo of the 1916 UK football team—all with concussions, probably—in the Margaret Ingels Collection).

Though you aren't able to copy and paste these images from your screen, it's simple to request copies of images you might use in a project—the procedures are set forth right there. These materials are there for researchers—which for the time being, at least, is what you are.

Beyond archives particular to the University and region, the **UK Libraries home page** provides links to numerous research sources, including these key ones:

- **InfoKat:** the main UK library catalog and search engine
- **LexisNexis:** the premier database for legal, news, public records, and business information
- The full-text **E-Journals Database:** containing entire articles from a range of scholarly journals
- **EBSCOhost:** a search engine for such journals, plus guides on these and a host of other databases, research tools, and search engines. Some, like LexisNexis, are available by institutional subscription only, thus accessible only from campus computers or through proxy servers for qualified off-campus users. Such powerful archival resources are expensive to subscribe to: it's a privilege of University life to have access at your fingertips.

Among the services available from the UK Libraries home page is **Ask-a-Librarian**—a link off of the Help section of the home page. There's a "chat" function open during business hours on weekdays, with a real librarian fielding instant-message questions; there are provisions for emailing or phoning in questions as well.

Linked to the Library website (just type it into their search line) you'll find **ARTstor** *(http://www.artstor.org),* a site run by the Mellon Foundation to provide art images and information for noncommercial, scholarly, and educational use. The site holds some 700,000 images at present, along with tools to help users locate and make use of them. Like LexisNexis and other subscription databases, ARTstor is most readily accessed through campus computers, though instructions are available for proxy server access by UK faculty, students, and staff off campus. This site is a remarkable trove of images, not just fine art but design, illustration, decoration of all sorts, even tattoo. It's a pleasure to surf, and its potential extends well beyond art history, to cultural phenomena of all sorts—a profusion of material for assignments in classes focusing on close readings of popular culture.

Here's a Sampling of E-Archives Beyond the UK Library's Links

- **The Library of Congress website** *(http://www.loc.gov)* contains a remarkable digital archive called **American Memory,** with texts, images, audio, and video of cultural artifacts in all sorts of areas. It lists these topics, with numerous links for each: advertising; African-American history; architecture, landscape; cities, towns; culture, folklife; environment, conservation; government, law; immigration, American expansion; literature; maps; Native American history; performing arts, music; presidents; religion; sports, recreation; technology, industry; war, military; women's history. It's all readily searchable and browsable—a national digital treasure.
- If you're working on an assignment about a place, you might go to **ePodunk** *(http://www.epodunk.com),* a website devoted, it says, to "the power of place."[11] It has information, maps, and links to further sites on some 46,000 communities across the United States, detailing parks, historic sites, museums, schools, cemeteries, and other features there and nearby. "Podunk" has long been a slang term for a generic tiny, obscure town; but there really is a Podunk, in upstate New York, and ePodunk's founders lived nearby, having retreated there from their former lives as journalists in order to create this site. It's a compendious labor of love, worth surveying.
- You may be enrolled in a class that takes citizenship and democracy as themes for exploration and research. A stimulating electronic source for such themes is **OpenDemocracy.net** *(https://opendemocracy.net),* described as "an online global magazine of politics and culture."[12] While this site is progressive in outlook (since democracy, after all, remains a progressive ideal), it is not partisan or doctrinaire but seeks, rather, to "publish clarifying debates which help people make up their own minds."[13] And it is indeed restlessly global in scope, providing wider, more varied coverage and perspectives on world affairs than we are accustomed to find in our news media. More than 1,500 articles are indexed at this site, representing writing on a range of themes (these categories are listed: arts & cultures; conflicts; democracy & power; ecology & place; women & power; faith & ideas; globalization; media & the net; people), from voices across the world, both celebrated and obscure. These voices may inform your own; you might even seek to join them, publishing your views on this site or others like it.

- A broadly useful electronic research source is **the Brookings Institution** *(http://www.brookings.edu)* website. The Brookings Institute is one of the best-known, most prestigious public affairs research institutes in the United States. The plethora of subject areas listed on its Research page indicates their great range of interests: business; cities and suburbs; defense; economics; education; environment and energy; governance; politics; science and technology; social policy. There are all sorts of leads for assignments like those common to writing classes. Some of our country's best minds, working on our most pressing issues, can be tapped in an instant here.
- Another prestigious research institute, conservative in outlook, is **the Hoover Institution** *(http://www.hoover.org),* located at Stanford University. Its website contains Library and Archives links useful for various projects. Other valuable information can be located through the site's Publications and Multimedia links. Publications include full texts of books and journals the Institution publishes. Multimedia includes access to *Uncommon Knowledge*, a weekly program of public policy discussion broadcast until 2005 on public television and since then as a webcast. The site offers full transcripts, streaming video, and downloadable MP3 files.

DOING ONLINE RESEARCH: WHY *WIKIPEDIA* IS NOT BAD FOR YOU

Increasingly, we tend to use online resources in our research process. Even professional writers and academic scholars use the Internet as part of their very serious research. Like anything else, the trick to using online sources is to use them *wisely.*

Wikipedia

I probably visit *Wikipedia* at least once a week. Maybe I am curious whether a celebrity is dead or alive. Or maybe I want to know everything I can about a certain holiday that is celebrated in another country. Or maybe I am trying to remember what year the first desktop computer was advertised. I have found answers to all of these questions on *Wikipedia.* I use it because I want quick answers to my questions.

Wikipedia is an ideal place to begin your research because the community of *Wikipedia* users regularly cite their sources of information. In fact, citation is one of the community standards that *Wikipedia* fans insist on. For this reason, you might begin by searching for a topic on *Wikipedia* and then scroll down to the bottom of the page to see what kinds of primary and secondary sources have been cited.

Let's say that you were interested in finding out the story of the great Kentucky meat shower of 1876. That's right. *The Kentucky meat shower.* According to *Wikipedia*, On March 3, 1876, chunks of raw meat fell from the sky for several minutes over Bath County, Kentucky.[14] Could it be that this strange event actually happened? Did Kentucky experience a meat storm over a century ago? If you scroll down to the bottom of the page, you will discover a footnote that sends you to a *New York Times* article from March 10, 1876 (**"Flesh Descending in a Shower: An Astounding Phenomenon in Kentucky—Fresh Meat Like Mutton or Venison Falling from a Clear Sky"** *[http://query.nytimes.com/mem/archive-free/pdf?res=9403E5D81E3FE73BBC4852DFB566838D669FDE]).*[15] If you find that *New York Times* article, you will learn that the meat was reported to be approximately the size of snowflakes and it reportedly tasted like mutton. It is certainly a strange story—and it makes for a great joke—but even in this strangest of *Wikipedia* entries, we can find reference to an important historical newspaper article. Even if the great Kentucky meat shower was a hoax, its 1876 nationwide reporting is worth noting.

Exercise: Using *Wikipedia* Wisely

Together with a group, choose a subject and look it up on *Wikipedia*. Your subject could be almost anything, but try to choose something that achieved some attention from a large number of people. (Examples: World Cup Soccer, emotions, home schooling, ADHD, President John F. Kennedy, etc.)

Read the *Wikipedia* entry and pay close attention to the footnotes. Choose one article that is cited in a footnote and try to find this source in the library's online databases.

Evaluating Online Sources for Credibility

Students often want to know how to evaluate the credibility of online sources. It's easy to Google your topic and find many kinds of websites that *seem* to address your topic. But are these websites the kind of sources that give you reliable information about your subject? If you have questions about whether or not online sources have strong *ethos*, ask yourself the following questions:

- What is the website or source of this material? Is this a website for a reputable organization? Is it a personal website? Look around the rest of the website and see if it contains questionable material.
- Is the source biased toward one particular opinion? Does the author or the website seem to have a particular goal in mind? If so, keep in mind that you are not getting an unbiased piece of information.
- Does the author have some kind of special qualifications or expertise on this subject? If the author is not listed, consider what it means that this piece is anonymous.
- When was this information published online? Is it outdated? Is newer information available?

VIDEO RESOURCES

Kevin Patterson on Doing "Research"

Kevin Patterson manages The Beer Trappe, a craft beer specialty shop in Lexington. Kevin is a native to eastern Kentucky, and he is an active participant in social media discussions of craft beer. As part of his job, Kevin uses rhetorical theory and writing in a number of ways. In this interview clip, he describes how he researches every day, as part of his job.

(https://vimeo.com/103939687)

NOTES

1 Sarah Altendorf, "074 | 365 August 11, 2011," *Flickr,* Uploaded on August 11, 2011, https://www.flickr.com/photos/46732441@N06/6033957917/.

2 "Castle + Versailles + Kentucky," *Google Search,* accessed June 27, 2014, https://www.google.com/webhp?tab=iw&ei=wPSxU4ffD8fOsATuiYG4Cg&ved=0CBcQ1S4#q=castle+versailles+kentucky&spell=1.

3 "Martin Castle," *Wikipedia,* last modified March 24, 2014, http://en.wikipedia.org/wiki/Martin_Castle.

4 Ibid.

5 Greg Kocher, "Woodford Votes to Approve Restaurant at CastlePost," *Lexington Herald-Leader,* last modified January 8, 2013, http://www.kentucky.com/2013/01/08/2468960/woodford-fiscal-court-votes-to.html.

6 Ibid.

7 Ibid.

8 "Viceroy, Filter the Smoke," *Flickr,* last modified July 6, 2007, https://www.flickr.com/photos/behindthesmoke/6012416034/.

9 Frank H. Jump, "Wrigley's Spearmint Gum—Vintage Print Ad 1911," Image, Uploaded March 27, 2012, http://www.fadingad.com/fadingadblog/2012/03/27/wrigleys-spearmint-gum-vintage-print-ad-1911-st-louis-fading-ad-1999/.

10 "Audio-Visual Archives," *University of Kentucky Libraries,* last modified September 19, 2013, http://libraries.uky.edu/libpage.php?lweb_id=391&llib_id=13.

11 "About ePodunk," *ePodunk,* last modified 2007, http://www.epodunk.com/about.html.

12 *OpenDemocracy,* accessed June 27, 2013, http://opendemocracy.net.

13 Ibid.

14 "Kentucky Meat Shower," *Wikipedia,* last modified April 25, 2014, http://en.wikipedia.org/wiki/Kentucky_meat_shower.

15 Ibid.

10 Chapter

Citation

WHAT IS CITATION, ANYWAY?

One of the most difficult things to do when writing is to cite.

What is citation? **Citation** is the quotation or paraphrasing of others' words, images, or ideas. Typically, we indicate citations by putting other writers' ideas and words into quotations and by indicating the page and source of the citation.

But why do writers cite each other's work?

Often, citation is the result of research. We've read a lot for a project, and we want to incorporate some of what we've read into our writing. To do so, we need to cite.

Figure 1. The punctuation marks we typically associate with citation.

But why do that? There are a number of reasons to cite:

1. **To give credit where credit is due:** This is the most repeated reason for citing. A writer gives credit to the original source in order to avoid being accused of plagiarism.
2. **Ethos (credibility):** Citation allows a writer to show that she knows the information at hand well. In fact, s/he knows it so well, s/he shows the ideas by citing them from their original sources. The reader will then find the writer to be credible (i.e., the writer isn't making all of this information up).
3. **Context:** Citation shows context. By context, we mean the larger picture. No idea exists in a vacuum. Ideas are part of larger conversations. Context shows where and how the writing being read fits within larger conversations.
4. **Reference:** Citations might note well-known ideas or make allusions to known people, sayings, events, moments, texts, etc. These citations can anchor the idea or writing in a reference so that the reader understands the writing's focus better.

5. **Service:** When a reader comes across an interesting idea in a given piece of writing, he may want to know where that idea came from so that he can read it completely at its source. A citation (and the accompanying Works Cited page) allows the reader to track down that source and read the idea in its original form.

There are two basic premises for including citations that we can note here:

- **In-Text Citation:** Referring to the source in the writing.
- **Works Cited:** Listing the writings, interviews, videos, films, songs, etc. used in the text when citing.

In-Text Citation
Jay Kesan notes that even though many companies now routinely monitor employees through electronic means, "there may exist less intrusive safeguards for employers" (293).

Entry in the List of Works Cited
Kesan, Jay P. "Cyber-Working or Cyber-Shirking? A First Principles Examination of Electronic Privacy in the Workplace." *Florida Law Review 54*.2 (2002): 289–332. Print.

Figure 2. Here's an example of both an in-text citation (top) and a Works Cited citation (bottom).

USING SOURCES

With these reasons for citing in mind, we might ask: what does a citation look like? Where do we get citations from?

What it shouldn't look like is a quotation lifted from an online citation site, such as **Quotations Page** *(http://www.quotationspage.com).*

And why not do this kind of citation? Because when you take a quotation from a quotation site, it is obvious that the quotation has been taken out of context, doesn't fit with the overall writing you are doing, has been jammed into your writing, and is not relevant. Instead of using a quotation site, cite from relevant material you have, material that shows you ideas, shapes your ideas, or performs any of the above items we've listed for citing. Cite from what you have:

- ◎ Read (books, magazines, journals, archival material, newspapers, magazines, websites)
- ◎ Seen (movies, videos, advertisements, posters)
- ◎ Heard (songs, interviews, podcasts)

Use the source for material. A citation page is not a source. Otherwise, your ethos (number 2 above) can be damaged. Readers won't believe you since they will know you lifted from citation websites and that you are not familiar with the source you are using.

This is where the difficulty comes into play. Many writers, when citing for the first time, struggle to figure out how to incorporate the citations into the body of writing. On the one hand, we want to use citations for the reasons outlined here. On the other hand, we don't want our citations to be awkward or merely jammed next to one another in a given paragraph or body of writing.

When citing from sources, we can first think of a number of roles citations play and then we can consider how to weave them together in order to write. That is, we don't want our citations to be an afterthought; we want them to be the writing itself.

Think of the weaving metaphor when citing in your writing. You are weaving, making fit, not jamming together quotations.

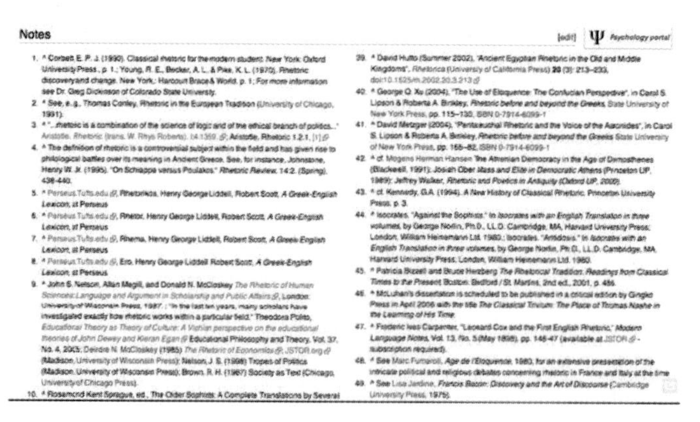

Notes [edit] Ψ Psychology portal

1. ^ Corbett, E. P. J. (1990). Classical rhetoric for the modern student. New York: Oxford University Press., p. 1.; Young, R. E., Becker, A. L., & Pike, K. L. (1970). Rhetoric: discovery and change. New York,: Harcourt Brace & World. p. 1; For more information see Dr. Greg Dickinson of Colorado State University.
2. ^ See, e.g., Thomas Conley, Rhetoric in the European Tradition (University of Chicago, 1991).
3. ^ "...rhetoric is a combination of the science of logic and of the ethical branch of politics..." Aristotle. Rhetoric (trans. W. Rhys Roberts). 1.4.1359; Aristotle, Rhetoric 1.2.1, [1]
4. ^ The definition of rhetoric is a controversial subject within the field and has given rise to philological battles over its meaning in Ancient Greece. See, for instance, Johnstone, Henry W. Jr. (1996). "On Schiappa versus Poulakos." Rhetoric Review. 14:2. (Spring). 438-440.
5. ^ Perseus.Tufts.edu, Rhetorikos, Henry George Liddell, Robert Scott, *A Greek-English Lexicon*, at Perseus
6. ^ Perseus.Tufts.edu, Rhetor, Henry George Liddell, Robert Scott, *A Greek-English Lexicon*, at Perseus
7. ^ Perseus.Tufts.edu, Rhema, Henry George Liddell, Robert Scott, *A Greek-English Lexicon*, at Perseus
8. ^ Perseus.Tufts.edu, Ero, Henry George Liddell, Robert Scott, *A Greek-English Lexicon*, at Perseus
9. ^ John S. Nelson, Allan Megill, and Donald N. McCloskey *The Rhetoric of Human Sciences: Language and Argument in Scholarship and Public Affairs*, London: University of Wisconsin Press, 1987. ; "In the last ten years, many scholars have investigated exactly how rhetoric works within a particular field." Theodora Polito, *Educational Theory as Theory of Culture: A Vichian perspective on the educational theories of John Dewey and Kieran Egan* Educational Philosophy and Theory, Vol. 37, No. 4, 2005; Deirdre N. McCloskey (1985) *The Rhetoric of Economics*, JSTOR.org (Madison, University of Wisconsin Press); Nelson, J. S. (1998) Tropes of Politics (Madison, University of Wisconsin Press); Brown, R. H. (1987) Society as Text (Chicago, University of Chicago Press).
10. ^ Rosamond Kent Sprague, ed., The Older Sophists: A Complete Translations by Several

39. ^ David Hutto (Summer 2002). "Ancient Egyptian Rhetoric in the Old and Middle Kingdoms". *Rhetorica* (University of California Press) **20** (3): 213–233. doi:10.1525/rh.2002.20.3.213
40. ^ George Q. Xu (2004). "The Use of Eloquence: The Confucian Perspective", in Carol S. Lipson & Roberta A. Binkley, *Rhetoric before and beyond the Greeks*. State University of New York Press, pp. 115–130, ISBN 0-7914-6099-1
41. ^ David Metzger (2004). "Pentateuchal Rhetoric and the Voice of the Aaronides", in Carol S. Lipson & Roberta A. Binkley, *Rhetoric before and beyond the Greeks*. State University of New York Press, pp. 165–82, ISBN 0-7914-6099-1
42. ^ cf. Mogens Herman Hansen The Athenian Democracy in the Age of Demosthenes (Blackwell, 1991); Josiah Ober *Mass and Elite in Democratic Athens* (Princeton UP, 1989); Jeffrey Walker, *Rhetoric and Poetics in Antiquity* (Oxford UP, 2000).
43. ^ cf. Kennedy, G.A. (1994). *A New History of Classical Rhetoric*. Princeton University Press, p. 3.
44. ^ Isocrates. "Against the Sophists." In *Isocrates with an English Translation in three volumes*, by George Norlin, Ph.D., LL.D. Cambridge, MA, Harvard University Press; London, William Heinemann Ltd. 1980.; Isocrates. "Antidosis." In *Isocrates with an English Translation in three volumes*, by George Norlin, Ph.D., LL.D. Cambridge, MA, Harvard University Press; London, William Heinemann Ltd. 1980.
45. ^ Patricia Bizzell and Bruce Herzberg *The Rhetorical Tradition: Readings from Classical Times to the Present* Boston: Bedford / St. Martins. 2nd ed., 2001, p. 486.
46. ^ McLuhan's dissertation is scheduled to be published in a critical edition by Gingko Press in April 2006 with the title *The Classical Trivium: The Place of Thomas Nashe in the Learning of His Time*.
47. ^ Frederic Ives Carpenter, "Leonard Cox and the First English Rhetoric," *Modern Language Notes*, Vol. 13, No. 5 (May 1898), pp. 146-47 (available at JSTOR - subscription required).
48. ^ See Marc Fumaroli, *Age de l'Eloquence*, 1980, for an extensive presentation of the intricate political and religious debates concerning rhetoric in France and Italy at the time
49. ^ See Lisa Jardine, *Francis Bacon: Discovery and the Art of Discourse* (Cambridge University Press, 1975).

Figure 3. Wikipedia entries are woven together by citations. Sources are listed by footnotes.[2]

Some online sources you are familiar with do a good job with citation. They do so in order to prove where their ideas come from. Contributors to *Wikipedia*, for instance, footnote each source they use for information. The citations build the encyclopedic entry for viewers. Readers of the entries can follow up with the sources if they want to read more. If a source is not cited, other contributors often demand citation as evidence.

Some of the examples in this chapter come from published work, but they don't have quotation marks nor do they have page numbers of their sources. The reason is because of style. In the style section of this chapter, we'll discuss how style determines the ways you cite. In the meantime, keep in mind that the usage of quotation marks and the indication of page numbers are not consistent throughout publications. In fact, many mainstream publications do not show their citations. This is a problem, of course. See above regarding service. Without page numbers, how can we follow up on the writer's ideas?

In the university, we do use quotation marks, indicate page numbers and sources within the writing itself, and include a Works Cited. You will do this kind of citation not only in WRD 110 and WRD 111, but in many other courses you take at the University of Kentucky. In academic writing—the bulk of writing you will likely do at the University of Kentucky—you'll use quotation marks, indicate page numbers and sources within the body of your writing, and include a Works Cited.

Sources for academic publications can be found in **UK's library** *(http://libraries.uky.edu).* In the Humanities, two major databases where you can find academic journals are JSTOR and Project Muse. They are accessible via UK's website.

ROLES OF CITATION

There are a number of ways we can use citations. Let's call these the "roles" of citation. The "roles" indicate different usages of a citation. Here are a few:

1. **Evidence:** The writer is making an argument. An argument needs evidence (proof) to demonstrate why it should be believed. The writer cites from statistics, examples, anecdotes, moments, common beliefs, or other forms of evidence by citing them from sources.

Example: To support a claim that films have become extremely violent in recent years, a writer might cite *Kill Bill's* The Bride versus Crazy 88 fight scene in which The Bride (Beatrix Kiddo) is able to single-handedly kill all of the members of the Crazy 88 gang over **an eight-minute time period.**

As evidence, the scene might be used to showcase the immense violence in a popular film, a film seen by millions of Americans.

Example: In sports, a controversial issue is the NBA rule that a player cannot be drafted until he is at least one year removed from high school. In 2012, former NBA basketball player and former Phoenix Suns general manager Steve Kerr **wrote an argument against removing the draft age limit** *(http://grantland.com/features/steve-kerr-problems-age-limit-nba/).* If a writer wanted to argue for the current rule, she could cite Kerr's positions on maturity, costs, player development, marketing, and mentoring as evidence. These five points could be cited as evidence a veteran of the NBA offers against removing the draft rule.

2. **Context:** One of the reasons we gave for citation was context. Context is also a role citation can play; it can provide a scene or setting for a piece of writing.

Example: You are writing about the struggle to be understood or to overcome obstacles in life (a fairly common experience). You need an example to show the context for what you've experienced (others have experienced it too, and it is important that the reader understand this point). You might cite Eminem's song "8 Mile" or the film *8 Mile,* which shows Rabbit's movement from the stamp factory to winning the rap battle. Rabbit overcomes all odds. You feel you've overcome all odds. Your experience has context, so you cite the film or song.

Example: In **"The Kentucky Derby Is Decadent and Depraved"** *(http://brianb.freeshell.org/a/kddd.pdf),* Kentucky's Hunter S. Thompson cites newspaper headlines in order to show the context of the time period he is writing about. The depraved Derby is put into context with a depraved time period of war. By reading the citations next to the Derby, the reader is supposed to read a larger context of depravity:

> *He grabbed my arm, urging me to have another, but I said I was overdue at the Press Club and hustled off to get my act together for the awful spectacle. At the airport newsstand I picked up a* Courier-Journal *and scanned the front page headlines: "Nixon Sends GI's into Cambodia to Hit Reds"... "B-52's Raid, then 20,000 GI's Advance 20 Miles"... "4,000 U.S. Troops Deployed Near Yale as Tension Grows Over Panther Protest." At the bottom of the page was a photo of Diane Crump, soon to become the first woman jockey ever to ride in the Kentucky Derby. The photographer had snapped her "stopping in the barn area to fondle her mount, Fathom." The rest of the paper was spotted with ugly war news and stories of "student unrest." There was no mention of any trouble brewing at a university in Ohio called Kent State.*[3]

diff. viewpoints

3. **Showing the conversation:** Often, writers need to show that a larger conversation exists, and that the writer is trying to be a part of that conversation. Citations show the conversation by weaving together relevant quotations (similar ideas, differing ideas) so that the reader sees how others have approached the topic.

Example: Grantland **writer Steven Hyden describes the importance of Led Zeppelin** *(http://grantland.com/features/the-winners-history-rock-roll-part-1-led-zeppelin/)* by showing a conversation among fellow writers Chuck Klosterman and Jon Landau as well as Jimmy Page. Notice how Hyden cites a little bit from each person in order to show a larger conversation regarding Led Zeppelin's sound. Hyden weaves together three sources to create the writing:

> *If it's still true (as Grantland's Chuck Klosterman once wrote) that every teenage male goes through a "Zeppelin phase," it is due to two things, and one of them is The Sound. Rock critic (and future Bruce Springsteen manager) Jon Landau once described The Sound upon seeing Led Zeppelin perform in 1969 as "loud … violent and often insane," but this is only two-thirds true as it pertains to Zeppelin's records. In reality, The Sound was the conscious and mentally competent creation of Jimmy Page in his role as Led Zeppelin's producer. Utilizing a coterie of well-chosen studio tricks—including something called "backward echo" and what he dubbed "the science of close-miking amps"—Page ensured that unlike practically all of his contemporaries in the late '60s and early '70s, his records would not be diluted by the passage of time.*[4]

Example: Writing about the demise of alternative media *(http://pitchfork.com/features/oped/9093-alt-weeklies/), Pitchfork's* Damon Krukowski weaves together several sources in one paragraph in order to show that alternative media often lost money:

> *In 1969, at the height of the underground press era, "72% of underground papers reported [making] no profit whatsoever," historian John McMillian points out in his book* Smoking Typewriters. *"Though they worked feverishly, most of them were jaundiced to the very idea of profit-making." Many operated as collectives without owners, some without editors. "There wasn't a hierarchical structure to what we were doing, so anybody could come in and get involved," says a contributor to Austin's The Rag in the colorful oral history of the 60s underground press, On the Ground. "All the underpinnings were different than they were in straight society," remembers one of the producers of Chicago's Seed. "There was a saying in the Seed, which I always believed: 'Work is love made visible.'"*[5]

4. **Definition:** When defining, the temptation is to turn to a dictionary and write something like "According to *Webster's* . . . ." But since anyone can open a dictionary and find a basic definition, a better approach is to cite from a text (written, filmic, visual) that shows the meaning you want to share. This way, you demonstrate for the reader that the definition exists in actual ideas (and not just in a dictionary).

Example: Malcolm Gladwell, **writing in "The Ketchup Conundrum"** *(http://www.newyorker.com/magazine/2004/09/06/the-ketchup-conundrum?currentPage=all),* defines ketchup accordingly:

> *Tomato ketchup is a nineteenth-century creation—the union of the English tradition of fruit and vegetable sauces and the growing American infatuation with the tomato. But what we know today as ketchup emerged out of a debate that raged in the first years of the last century over benzoate, a preservative widely used in late-nineteenth-century condiments.*[6]

Gladwell defines ketchup by drawing on a historical citation. He doesn't show us where he got it from, but we can assume that this is not common knowledge he has in his head; he must have read it somewhere.

5. **Appeal:** In order to persuade a reader to believe in a given idea, a writer may want to show an emotional example. The writer may, then, appeal to the reader's emotions or beliefs by citing a source.

Example: In 2012, **a video showing "pink slime"** *(https://www.youtube.com/watch?v=d-CqKl4Q3hW4),* meat processed mostly for fast food consumption circulated on the Internet. By showing fast food meat as a pink sludge of animal parts, the video meant to appeal to people's distaste for food that looks and sounds unappetizing. In turn, it meant to dissuade people from eating fast food.

6. **Balance:** Some writers want to navigate an issue by citing from differing and conflicting positions so that a pro/con argument is not presented. Instead, various sides are shown in order to present a more complex response than the limited pro/con binary.

Example: Reporting on the "pink slime" controversy associated with fast food and processed meat, **the website Snopes** *(http://www.snopes.com/food/prepare/msm.asp)* cites a number of published documents in order to navigate "a mixture of true and false information."[7]

Thus, the citations provided balance to an otherwise controversial idea.

7. **Repetition:** Often, ideas repeat in a number of contexts. A writer may feel that showing a specific repetition will produce a specific effect for the reader. Therefore, the writer cites the repetitions.

Example: The *Daily Show* introduced Barack Obama's 2009 inaugural speech by showing it was a repetition of George Bush's inaugural speech. The citation of a repetition was meant to show little difference between the two presidents, despite a general public's feelings that they had elected a representative of change.

Example: Musical artists often use repetition to cite other songs, muscially or lyrically. The songs may not explicitly state the origin of what is repeated, but they repeat an idea to create new meaning .

Go on **YouTube** *(https://www.youtube.com)* and look up the following repetitive usages of "Try a Little Tenderness" by the following artists:

- Bing Crosby
- Solomon Burke
- Aretha Franklin
- Otis Redding
- Tom Jones
- Kanye West and Jay-Z

How does each artist repeat the basic elements of the original? How do the meanings change in each new context?

8. **Introductions:** A writer may want to introduce an idea by citing another source. Such citations can be better beginnings to an idea than clichés (*"Since the beginning of time …")* or vague generalities (*"A lot of people think different ideas about this topic …"*). The citation can offer a specific entry into the idea you want to convey.

Example: In a 2013 essay on Syracuse University's men's basketball team's run for the Final Four *(http://www.theatlantic.com/entertainment/archive/2013/03/sorry-syracuse-why-the-hot-hand-in-basketball-maybe-isnt-a-real-thing/274489/),* the *Atlantic's* Ashley Fetters introduces the concept of the "hot hand" by citing a 1991 study. While Fetters paraphrases her source, she does cite the concept ("hot hand"), the authors (Gilovich, Tversky, and Vallone), and the source (the *Wilson Quarterly*). This citation is meant to introduce readers to the concept in order to argue against the idea of a "hot hand":

> *In 1991, psychologist Thomas D. Gilovich and his colleagues Amos Tversky and Bob Vallone conducted the classic investigation of the phenomenon of the "hot hand"—that is, when a player gathers shooting momentum or gets "in the zone" to the extent that he or she seemingly can't miss a shot—and published their findings in* The Wilson Quarterly. *They discovered it didn't exist anywhere outside of players' and fans' imaginations; rather, the research team found, the "hot hand" happens when an optimistic interpretation and some funky psychology are applied to a statistically average outcome of shots missed and shots made.*[8]

WEAVING

To be effective, citations have to be the writing itself. That is, the citations are the writing, not added to the writing. Thus, we introduced the metaphor of weaving in order to better express how to work citations into your writing. You don't drop them in awkwardly; instead, you weave them into the writing so that they are the writing.

Sometimes students think: *"I'll just throw in a citation here or a citation there."* But that kind of approach comes off as awkward.

If you follow up on the preceding examples, you'll see that these citations make up the body of the writing. The writers don't add citations as if they were seasoning a meal; these writers build a piece of writing out of the information that they have researched.

In a 2013 *New Yorker* piece on Guantánamo Bay *(http://www.newyorker.com/news/daily-comment/a-hundred-hungry-men-at-guantnamo),* Amy Davidson weaves together a question asked of President Obama at a press conference and his answer:

> *"Is it any surprise, really, that they would prefer death rather than have no end in sight to their confinement?" a reporter asked during a White House press conference on Tuesday. President Obama was aware: "It is not a surprise to me that we've got problems in Guantánamo," he said, and then talked about how he said so back in 2007, and how that was part of his campaign, and how Congressional Republicans had made things very hard. They have; but he, for a couple of years now, has made it easy for them to do so. He said he knew that it wasn't "sustainable."*[9]

In a 2013 *Atlantic Monthly* article *(http://www.theatlantic.com/national/archive/2013/04/will-new-teacher-evaluations-help-or-hurt-chicagos-schools/275415/)* on teacher evaluations in Chicago public schools, Sara Neufeld weaves a statistic from the *New York Times* with a quotation from an expert:

> *The* New York Times *recently reported that only 3% or fewer of teachers under new systems in Florida, Tennessee, and Michigan were rated unsatisfactory, raising questions about how effective that agenda will be. "The hope of the business folks is that this ... will identify more failing teachers so they'll be gotten rid of. I'm just not sure that's going to happen," said Sue Sporte, research operations director at the University of Chicago's Consortium on Chicago School Research, which is studying the implementation of the evaluations in the Windy City. "If only 2.5% is rated unsatisfactory, I don't know how satisfactory that will be to the constituency that believes that kids are failing because teachers are awful."*[10]

When you read—whether in popular publications such as the *New Yorker* and the *Atlantic Monthly* or in academic publications that you might find via the library—notice how authors weave together their citations. How do they create a fluid writing from citations?

CITATION STYLES

It's too easy to get hung up on **style.** Citation style mainly means the ways you note the citation, where you put the page number and source, and how you construct a Works Cited. Styles aren't holy. They are discipline/profession specific. Styles allow us to format our citations in a way that are readable to a specific audience. If you write in the university, you are often expected to use a specific citation style, depending on your audience.

You've likely noticed that a great deal of what you read outside the classroom—newspapers, websites, novels, magazines—often lacks stylistic citations. That is, even if the writer conducted research for the writing, there often isn't a Works Cited or even a page number and source. This is also true for many of the examples presented in this chapter.

The citation styles are largely academic. They are professional ways of indicating sources. Thus, when you use a citation style, you are engaging with the professional way of doing something. Every profession has its own ways of presenting ideas, documents, and sources. In the university, the professional way of citing sources is to use one of the academic citation styles.

Style is the easiest part of citation. Style is merely the way you format and present your citation. It does not reflect the quality of your citation. How do you know which style to use? The key is knowing who your audience is. Each style is based on an academic audience:

- **MLA:** mostly humanities (English, foreign languages)
- **Chicago:** history
- **APA:** mostly sciences (psychology, physics, biology, technical writing)

Nobody can memorize the styles for citing books, movies, articles, interviews, etc. And there are enough online resources that make this information available that you can find the proper style easily.

Some Style Resources

- **Purdue OWL** *(https://owl.english.purdue.edu):* all of the styles are documented here.
- **Zotero** *(https://www.zotero.org):* a free tool that allows you to gather online sources.
- **MIT Overview of Citation Software** *(http://libguides.mit.edu/references):* an overview of four major software packages.

To practice working with citations, here are a couple of exercises. The exercises are meant to give you some familiarity with identifying and working with citations.

Exercise: Reverse Citation

The reverse citation exercise asks you to take a piece of writing that has citations—the writer obviously has used outside sources to write—but doesn't indicate where the citations came from. The reason for doing the exercise is to try and understand how writers use citations in their work.

This exercise asks you to find the citations that likely informed the writer.

Using the following essay, **"Dissed Fish" by Calvin Trillin** *(http://www.newyorker.com/magazine/2004/09/06/dissed-fish?currentPage=),* identify places where Trillin is likely using an outside source. You can tell if he is using an outside source by asking yourself: *"Could someone know that on his/her own?"* If the answer is "no," then you need to find the source.

Use the following tools to guess where the writer got his information from. The keyword here is *guess*. We don't really know if the writer used these texts. But the guess will allow us to figure out how a writer might find a source and then use information from that source in his/her own writing. Here are some places where you might want to start your search:

- **Google Books** *(https://books.google.com/?hl=en)*
- **Google** *(https://www.google.com/?gws_rd=ssl)*
- **Magportal**
- **UK's Library portal**

Make a list of the sources you find. Then compare them to how Trillin uses the information to write "Dissed Fish." Did you find the exact source (i.e., what you found matches his writing) or something that might be the source (the same idea)?

Exercise: Citation Profile

This exercise asks you to build a profile of a public figure by researching and then citing sources related to that figure. The citations you find will show a reader who that person is. You are not writing a biography as much as you are building a profile via citations. A public figure may be one of the following:

- A celebrity
- A politician
- An artist
- Someone else in the public eye
- An entertainer
- Activist
- A business person

The first thing you need to do is identify a person. Then, research as much as you can about that person on the following:

- Their beliefs
- Hobbies
- Background
- Other's views on that person
- What he/she has done
- Habits
- Successes or failures

Use magazines, newspapers, public records, interviews, online books, websites, and other similar sources in order to find information for your profile.

With each source you find, identify a few passages that demonstrate that point. For instance, if you were reading the ***New Yorker's*** **profile of Whole Foods CEO John MacKey** *(http://www.newyorker.com/magazine/2010/01/04/food-fighter),* you would see that MacKey is interested in the following:

- Yoga
- Veganism
- Libertarianism
- The writings of Ayn Rand

If you were doing a citation profile of John MacKey, these could have been four items you researched and found citations about.

Your next step would be to compose a paragraph or two about John MacKey from the citations.

You could have multiple citations for each item. By weaving the citations together, you will be able to build a profile. The reader will know more about the person you chose to profile by the strength of your citation choices.

NOTES

1 *MLA Research Paper FAQs*, accessed June 27, 2014, http://mrssperry.com/MLA%20 Paper%20FAQs%20 Table.htm.

2 "Rhetoric," *Wikipedia*, last modified June 13, 2014, http://en.wikipedia.org/wiki/Rhetoric.

3 Hunter S. Thompson, "The Kentucky Derby is Decadent and Depraved," *Scanlon's Monthly 1,* no. 4 (1970).

4 Steven Hyland, "The Winners' History of Rock and Roll, Part 1: Led Zeppelin," *Grantland*, January 8, 2013, http://www.grantland.com/story/_/id/8821559/the-winners-history-rock-roll-part-1-led-zeppelin.

5 Damon Krukowski, "Alternate Ending," *Pitchfork*, March 27, 2013, http://pitchfork.com/features/oped/9093-alt-weeklies.

6 Malcolm Gladwell, "The Ketchup Conundrum," *New Yorker*, September 6, 2004, http://www.newyorker.com/archive/2004/09/06/040906fa_fact_gladwell?currentPage=all.

7 "Pink Slime and Mechanically Separated Chicken," *Snopes*, last modified March 22, 2012, http://www.snopes.com/food/prepare/msm.asp.

8 Ashley Fetters, "Sorry, Syracuse: Why the 'Hot Hand' in Basketball (Maybe) Isn't a Real Thing," *Atlantic*, March 29, 2013, http://www.theatlantic.com/entertainment/ archive/2013/03/sorry-syracuse-why-the-hot-hand-in-basketball-maybe-isnt-a-real-thing/274489/.

9 Amy Davidson, "The Guantánamo Hunger Strikes: An Unprecedented Hundred Hungry Men," *New Yorker*, May 1, 2013, http://www.newyorker.com/online/blogs/ comment/2013/05/guantanamo-hunger-strikes-hundred-hungry-men.html.

10 Sara Neufeld, "Will New Teacher Evaluations Help or Hurt Chicago's Schools?" *Atlantic Monthly*, April 30, 2013, http://www.theatlantic.com/national/archive/2013/04/will-new-teacher-evaluations-help-or-hurt-chicagos-schools/275415/.

Writing with Style
11
Chapter

You have an implicit understanding of rhetoric: You know that you must adapt your language to suit your audience and the particular occasion. If you want to explain to your friends why you didn't like a movie, you will make one argument; if you want to contribute to a discussion of the same movie in your film theory class, you will make a different one. Your conclusion—that the movie was a poor one—will likely remain consistent. But you might choose to highlight different elements of the movie—and you will almost certainly use a different vocabulary and tone in making your point to these two different groups. These decisions you make about word choice and tone are stylistic ones, and they are every bit as rhetorical and important to your success in convincing your audience as the choices you make about content.

THE APPEAL OF STYLE

Although stylistic choices can make a piece of writing "sound good," an effective style is not just a beautiful one. Style also effects an audience's perception of a text's reasonableness and the character of the writer. One way to understand the impact of style is to consider it in terms of the three primary rhetorical appeals:

- **Logos:** An appeal based on good reasons (logic)
- **Pathos:** An appeal to the emotions or to values (personal and cultural)
- **Ethos:** An appeal based on the character (authority and credibility) of the writer

In order to be successful, a writer must construct a text that appeals to readers through their logic, their emotions, and their sense of the writer's character. And an impactful style (diction, voice, syntax, etc.) is necessary to accomplish all of these ends. Put more specifically, an effective style makes an argument easier to follow (makes a logos appeal); it creates a mood and a sense of aesthetic pleasure (makes a pathos appeal); and it convinces the audience the writer is knowledgeable and trustworthy (makes an appeal to ethos).

If the stylistic choices of the writer are ineffective, a text can fail to win over the audience regardless of its content; the writer may seem unkind, or the mood may feel inappropriate. Consider, for example, this piece by Prachi Gupta for *Salon.com* on "The 15 Most Hated Bands of the Last 30 Years." *(http://www.salon.com/2013/08/10/15_most_hated_bands_of_last_30_years/)* Gupta chooses a biting, humorous tone, as in her justification for John Mayer's inclusion on the list:

> *John Mayer is that insufferable bro—you know, the one who wears a pukka bead necklace, is always shirtless, toting around a guitar at that house party you didn't want to go to, anyway. He'll suck the humor out of a joke and ruin the punch line every time, but no one else seems to care, because he's a shirtless bro with a guitar. He probably likes Dane Cook. And misogyny.*

Gupta might describe her ethos, her persona for the piece, as trendy and humorous, but the comments below the piece suggest that much of her audience would have described it differently. Few commenters disagreed significantly with Gupta's choices (which included Creed, Nickleback, Limp Bizkit, and 98 Degrees), although they often added to the list. But the article nevertheless aroused a good deal of vitriol. One commenter, Local Area Man, claimed, "I feel like I'm reading an eyeball rolling, impatient teenager who is bored and angry." Tejana, another commenter, scoffed, "Who elected you the taste Czar?"

For many of her readers, Gupta creates a dislikeable persona through her writing: She fails to build an effective ethos. More importantly, we could point to elements of her text that led her readers to describe the piece as "insipid" and "self-indulgent." The repetition of the word "bro" and short sentences at the end make the paragraph seem not just humorous but snarky—humor at the expense of a group of people (those who attend house parties, use the word "bro," and play guitar for others unsolicited). To further associate this group with misogyny seems both unnecessary and unfair. She also uses the second person, "you," suggesting that the reader must automatically share these opinions.

Gupta's missteps in this article are mistakes less of content than of style—of tone and word choice. You could make an argument that her piece is "link bait" or "troll bait"—deliberately antagonist in order to entice people to repost the link or comment. However, those readers who found her ethos to be off-putting are unlikely to click on Gupta's articles in the future. Any short-term success may mean long-term failure.

No one style will be effective for all contexts, let alone effective for all audiences. Instead, the key to success is learning to make informed stylistic choices that will enhance the content of your writing. This chapter will help you make choices that will meet the needs of your text and your audience's expectations.

VOICE

I have used the word ethos to describe Gupta's writerly persona in her piece. When we judge a writer's ethos, we consider both the content and stylistic features of the text: our sense of how informed they are; our ability to identify culturally and personally with their persona; and their **voice,** which is most deeply connected to style.

You likely already have a sense of what the word voice means in writing—the personality of the writer as it comes through on the page. But we can be much more specific in analyzing what makes an impactful and appropriate voice if we consider the stylistic elements that create it—particularly its perspective, tone and diction, and use of figurative language.

Perspective

Perspective, or point of view, is the stance that you take up to address your topic. Do you want your position to come across as very personal (first person)? Objective or impersonal (third person)? Or you want the piece to seem conversational, by directly addressing the reader (second person)?

First Person	Second Person	Third Person
I, we (my, me, us, our)	you (your)	he, she, they, it (his, her, their, him, her, them, its)

The table above lists the pronouns associated with each perspective. The perspective you choose should depend on your topic, the particular context of the piece, and your sense of the ethos you want to develop. For example, Deb Perelman uses a first person perspective in her blog, *Smitten Kitchen:*

> *Look, I came around on kale, I did! I realized that I didn't dislike* it *so much as much as I was suspicious of fervor around it, as if there had never been any other healthy vegetables before it, as if it its renaissance was the result of the kind of PR team only a certain troubled mayoral candidate could dream of right now. And the kale chips of 2010 didn't do much to convince me (ducks; sends self home from the internet in disgrace). But it turns out, I like kale the way I like my slaws—raw, finely slivered, not overwhelmed by dressing and with just the right extra punches to round it out. ("Kale Salad with Pecorino and Walnuts";* http://smittenkitchen.com/*)*

The first-person perspective is standard for the blog genre. It helps the audience to identify with Perelman as a lover of food, rather than a chef (amateur or not). This approach allows Perelman to portray cooking not as an exact science but as an experience that the audience can share from their own kitchen—from picking the recipe and ingredients to preparing the dish. We move with her through the process of culinary creation, thanks largely to her first-person stance.

Tone and Diction

Perspective, however, is only the first decision you'll need to make in developing a powerful voice for a piece of writing. The stylistic elements of a text that will be most important in making your voice appropriate or distinctive are your tone and diction. **Tone** reflects a stance, attitude, or emotional investment in a topic. **Diction** is word choice, which helps to build a particular tone.

Your tone and diction should supplement your choice in perspective and be appropriate for the content and rhetorical situation. Business reports, for example, are usually written in third person, with a tone that is straightforward rather than humorous, cheerful, or angry. Blogs, like Perelman's above, often supplement their first-person perspective with an informal diction and an intimate or funny tone.

But there's more to creating the right tone than simply labeling it with an adjective. Let's compare Perelman's voice to that of Megan Carter in another food-related blog, *Verdant Nation:*

> *[H]ow can cooking our own meals help the obesity epidemic? Meals and snacks eaten outside of the home generally have more calories than those made in the home. Simplistically, if you consume more calories than you expend on a regular basis, you're going to gain weight. There is also some evidence that eating more frequently outside of the home is related to an increased body weight. Preparing your own meals also cuts down on packaging, particularly if you eat a lot of fast food, which is better for the environment. And preparing your own meals means just that—as Yoni Freedhoff recently commented, nuking something doesn't count. ("Is Cooking the Silver Bullet to the Obesity Epidemic?"* http://verdantnation.blogspot.com/)

The first thing that we notice is that Carter's paragraph uses a third-person perspective, as opposed to Perelman's first-person. (The "you" in this case is less a direct address to the individual than a synonym for the overly formal "one" or "an individual.") This choice in perspective gives Carter's piece an analytical, authoritative tone, rather than the intimate, playful tone of Perelman's. The perspective of each piece suits its purpose: Perelman hopes to make the art of cooking a part of her readers' everyday lives; Carter aims to relay scientific research on food and obesity to a non-scientific audience and reflect on its significance.

But we could highlight further stylistic elements that distinguish the tones of these blogs. Perelman's piece includes an exclamation, a parenthetical aside, and interjectory words ("look," "well"). All of these elements give the piece its more informal feel. By contrast,

Carter's piece uses a standard argumentative technique: Ask a question in the first sentence, then answer it. It does not include the stylistic features that Perelman's does. Instead, it opens sentences with transitional adverbs ("simplistically") and uses abstract subjects ("there is"; "preparing your own meals"). These choices create a more authoritative, formal tone.

The diction of each piece further defines the voice. Certainly, both blogs use standard or everyday diction, if by standard we mean non-specialized. But if we look carefully at each paragraph, we can see more subtle differences. Perelman frequently uses phrasal constructions, where multiple words are used to convey a single idea ("came around"; "turns out"; "round out"). And, of course, she makes jokes and uses slang ("PR team"; "kale chips of 2010"; "sends self home"). Because these are features of quite informal texts, Carter uses them less frequently in discussing scientific topics. Perspective, tone, and diction work together in both texts to create a voice that is coherent and appropriate.

Figurative and Descriptive Language

Perhaps the most imposing element of creating a strong voice is using figurative and descriptive language. **Figurative language** is language that compares two things in a non-literal (and hopefully unique) way. **Descriptive language** creates a mood or paints a picture by playing on the audience's senses.

Sam Sifton draws his audience into his discussion of summer cooking with both figurative and descriptive language:

> *Peak summer is upon us, the corn as high as an elephant's eye, tomatoes fat and nearly bursting on the vine. On the East Coast, down in the shallows of low tide, you'll find people doing the shimmy with their bare toes, or pulling hard with a rake, looking for clams. There is basil in the window box and bacon in the fridge. This is the time of year to make like a chef: Embrace the simplicity of this, and allow the ingredients to talk for themselves.* (http://www.nytimes.com/2013/08/18/magazine/for-summer-cooking-embrace-simplicity.html?ref=dining&_r=0)

The piece opens with a familiar simile ("high as an elephant's eye"). While such a stock simile would not normally be successful, in this case it makes the image feel universal. Then, he quickly progresses to a series of images of summertime using rich descriptive language. He invokes particularly touch ("bare toes"; "pulling hard with a rake") and sight ("basil in the window box and bacon in the fridge"). This series of images creates a particular mood, a sense of nostalgia, from the opening lines of the piece.

While using figurative and descriptive language may sound difficult, you use both all the time. You use metaphor ("a breath of fresh air"); simile ("strong as an ox"); hyperbole ("it took an eternity"); and even metonymy ("the White House" to stand for the presidential administration) and synecdoche ("wheels" to stand for a car). These figures are already a part of your everyday discourse. Creating new figures involves nothing more than making new and unusual comparisons: these new creations will help you continue to develop your own unique voice.

It would be impossible to discuss all of the many tropes and figures available to you in the confines of this chapter. However, the table below lists some common tropes that will be useful to you, along with a definition and an example.

Figure	Definition	Example
Metaphor	Describing one object using the characteristics of another.	The sunflowers bobbed their heads in the strong Midwest breeze.
Simile	Comparing two things using "like" or "as."	The creek bustled with the same energy as a busy city street.
Alliteration	Repeating the same sound in nearby words.	The lake lapped against my legs, lulling me to sleep.
Metonymy	Referring to something by naming an object associated with it.	The Crown issued a proclamation when the new heir was born.
Synecdoche	Referring to an object by naming one of its parts or attributes.	We take only cash or plastic.
Anaphora	Repetition of the same word(s) at the beginning of consecutive clauses or sentences.	I came; I saw; I conquered.
Traductio	Repeating the same word in various points of a sentence.	Desperate times call for desperate measures.
Anastrophe	Shifting normal word order to create effect and emphasis.	Against the vast blackness hung a brilliant full moon.

You can find many more tropes and figures using online resources; Silva Rhetoricae is particularly useful in this respect *(http://rhetoric.byu.edu/).*

Exercise

Choose a paragraph from an article or blog that you enjoyed. Rewrite the paragraph, keeping the content but re-imagining the voice. In other words, change the tone, diction, and perspective so that is differs from the original. It can help to imagine an entirely new context for the paragraph. If the paragraph came from a magazine, perhaps imagine it as an angry letter to the editor; if it came from a blog, imagine it as a front-page news article or a brochure.

SYNTAX

If a strong writerly ethos depends on developing an appropriate voice, then a successful logos appeal, or appeal to reason, is grounded in an effective use of syntax. **Syntax,** or sentence structure, involves not just using correct grammar—but building sentences that accurately reflect the meaning that you want to convey.

As you read through this section on syntax, it is important to remember that there is no one right way to build a sentence. Any standards of "good" and "bad" English have varied significantly over time. English-speakers during the Renaissance highly valued a writer's ability to use tropes and figures and applauded what we would call flowery language. By contrast, contemporary English-speakers (particularly denizens of corporate America) value efficiency in prose, or the "skimmability" of a piece of writing. These two standards call for different types of syntax.

However, just because they are many ways to write a good sentence doesn't mean that all sentences will be equally effective. First, different types of sentences signal different kinds of relationships between ideas, so some sentences will better reflect what you mean than others. Second, you have an obligation to your readers to make your writing clear. They are investing time in your writing, so it's your job to make their reading experience as smooth and pleasurable as possible. The table below outlines four general principles—four "rules of thumb"—for using syntax effectively.

The Four Principles of Syntax

1. English speakers prefer simple, concrete subjects.
2. English is a verb-driven language: The verb of the sentence should always be in the simplest tense that is appropriate.
3. Subjects and verbs should be placed close together and only occasionally separated by modifying phrases and clauses.
4. English sentences (usually) branch to the right, toward the end the sentence. The simplest and most concrete elements of the sentence should be at the beginning, while complex modifiers are added at the end.

This section will discuss these principles in more detail and give you strategies for implementing them in your own writing. Great writers are those who develop a unique and impactful voice, while abiding by the principles of effective syntax; it is these writers that both engage their audience and meet their expectations.

Choosing Effective Subjects

Principle 1 of the Principles of Syntax states that "English speakers prefer simple, concrete subjects." In other words, sentences are easier to read when they begin with subjects that can actually *do* something—or that we can at least see: I, my mother, the ball, the university administration, etc. These types of subjects are easily followed by simple action verbs: I wrote; my mother left; the ball rolled; the university administration attempted, etc.

However, all writers sometimes allow their sentences to be taken over by relatively "empty" subjects. The most common empty subjects are found in Table 4, along with their most common verbs. Sentences that begin with these expressions have a filler for a subject, such as "this," "it," or "there," rather than a real subject that can do something. These expressions are syntactical fillers: They have no meaning but merely give the sentence a grammatical subject and verb.

Empty Subject and Their Verbs

- ◎ It is/was
- ◎ This is/was
- ◎ There is/are
- ◎ It seems
- ◎ I think
- ◎ I believe

Writers tend to use these expressions because they occur in conversation. But because we value efficiency so highly in writing, these same subjects cause problems when they are put on paper. First, English is a verb-driven language, and strong verbs require strong subjects. A vague subject like "it" often leads to a weak verb like "is." Second, they make a sentence wordy and complicate the sentence structure. An expression like "this is" at the beginning of a sentence merely fills in for the actual actor of the sentence—the real subject. The real subject then gets buried further along in the sentence, making it harder to read and leading to grammatical errors.

Writing Strategy

One effective means of revising at the sentence level is to focus on empty subjects. Try rewriting the same problematic sentence with two or three different subjects. Which sounds best? Choosing new subjects is just an easy way of brainstorming different ways to state an idea: more than likely, one of the sentences on your list will work for you.

Consider, for example, this sentence:

- ◎ **There are great resources available to student organizations when it comes to facilitators and presenters.**

The sentence begins with the empty words, "there are," and more importantly, the sentence is unclear and vague. So we want to give it a better subject that more clearly expresses the idea.

Writing Strategy (continued)

The subject of the sentence should be most important actor in the sentence—the person or thing that *does* something. The easiest way to find the "actor" of a sentence is to consider its most important nouns. In this sentence, the most important nouns are resources, organizations, facilitators and presenters. With this in mind, we can begin crafting alternative sentences, based on our new subjects:

- **Facilitators and presenters can serve as important resources for student organizations.**
- **Student organizations can plan presentations to meet the academic or professional needs of its members.**

We might also have a "hidden" subject that doesn't appear in the original sentence:

- **The Office of Student Affairs can help student organizations find facilitators and presenters that will appeal to their members.**

Obviously, these sentences say different things, and the writer would have to choose the one that best expresses his or her meaning. But all of these sentences are clearer and more specific than the original, largely because they have a strong subject.

You don't need to rewrite every sentence multiple ways—just those that seem awkward. Best of all, you do not always need to know *why* a sentence isn't working in order to improve it. Oftentimes, these awkward sentences don't even have grammatical mistakes—they are just wordy, and a better subject will eliminate the problem indirectly.

Choosing Impactful Verbs

Principle 2 of the Principles of Syntax states that "English is a verb-driven language: The verb of the sentence should always be in the simplest tense that is appropriate." In other words, your verbs should, if possible, be in simple present, past, or future tense: jumps; jumped; will jump. Certainly, you frequently will need other verb tenses, such as progressive forms (was jumping), to express certain ideas, but you should be selective in your use of them. Sentences are clearer and often richer with strong, simple-tense action verbs.

This section addresses two common writing problems that may impede your ability to build sentences with strong, impactful verbs: passive voice and nominalizations.

Passive Voice

Undoubtedly, you've had a teacher warn you of the dangers of passive voice. Passive voice occurs when the real object of a sentence becomes the subject, and a simple action verb is replaced by a form of the verb "to be" and a participle. Passive voice inevitably causes the writer to break Rule 2: "wrote" becomes "was written," and "cook" becomes "was cooked."

- **Active:** My ball broke the window.
  Subject Action Verb Object
- **Passive:** The window was broken by my ball.
  Subject "to be"+ participle prepositional phrase with the real actor

As you can see in the example above, the passive voice sentence, although expressing the same idea, is wordier and has a more complex verb. Moreover, it buries the real actor of the sentence: "The ball" is hidden in a prepositional phrase at the end of the sentences, rather than holding its rightful place as the subject. The window, real object of the sentence—the thing being acted upon—fills in as the subject, where it does not belong. When you are trying to express a more complicated idea than the one above, this convoluted structure causes even more problems.

A sentence will almost always read more clearly if the main actor in the sentence is the subject. However, like most rules of writing, there are exceptions to the rule that we avoid passive voice. The passive voice may more appropriate when:

1. The passive construction simplifies the subject of the sentence.

   For example, consider one of the clauses that I wrote above:

   - A simple action verb is replaced by a form of the verb "to be" and a participle.

   This verb of this sentence—"is replaced"—is in passive construction; it includes a form of the verb "to be" and a past participle, "replaced." I could have written the sentence in active voice:

   - A form of the verb "to be" and a participle replaced a simple action verb.

   This version of the sentence has a simple past-tense verb, "replaced," but it also has a very long subject: "a form of the verb 'to be' and a participle." I decided that the passive construction was a "lesser evil" than the overly long subject. In other words, a simple subject takes precedence over an active verb.

2. The audience expects passive construction as a convention.

   In some cases, an audience will expect a passive construction, whether it simplifies the sentence or not. For example, scientists will always use a passive construction when describing their experimental method. They will not say, "I diluted the solution," but "the solution was diluted." This structure buries the true actor ("I") but does so for rhetorical and ideological reasons. The passive construction implies that it does not matter who dilutes the solution—the results should remain the same regardless: it implies that the scientific process is "objective" and the results "true."

Nominalizations

While passive voice is a common problem, we often have to look beyond the main verb of the sentence when we are revising our prose. Nouns are a very flexible part of speech, one that can often mutate into other forms: In particular, both verbs and adjectives can often be turned into nouns. These words are called **nominalizations.**

Writers often use nominalizations in an effort to "sound smart" or more formal because they are more abstract. But more often than not these words just confuse the primary action happening in the sentence. Consider the following sentence:

> *While the meta-***analysis** *will yield a mathematically sound* **synthesis** *of the studies included in the* **analysis,** *if these studies are a biased sample of all possible studies, then the mean* **effect** *reported by the meta-***analysis** *will reflect this* **bias.** (https://www.meta-analysis.com/downloads/Intro_Criticisms_optim.pdf)

This short sentence includes several nominalizations, bolded in the sentence above: analysis could be analyze; synthesis could be synthesize; effect could be affect; bias (n.) could be bias (v.) or biased (adj.). These nominalizations make the passage sound sufficiently "scholarly" but make it far more difficult for the audience to understand.

Writing Strategy

The English language contains so many nominalizations that you will use them regularly. Rather than avoiding them completely:

1. Primarily avoid using nominalizations as the actor of the sentence or clause.
2. Identify and eliminate nominalizations as a means for revising problematic sentences.

In the sentence above, the main subject is "the mean effect," a nominalization. Rather than using effect as a noun, we might consider revising our sentence by beginning with the verb affect.

Next, we must choose a subject for our new sentence. What is affecting the analysis? In this case, it is a biased sample. Now we have the beginning of a solid sentence: **"A biased sample will affect ..."** We can continue to build the sentence from this foundation:

◎ **A biased sample of studies will affect the results of the meta-analysis, even if it accurately synthesizes those studies included by the author.**

The sentence above eliminates most of the nominalizations, although "meta-analysis" remains. More importantly, it conveys the same information to the reader in a much clearer way.

Building an Effective Sentence

To this point, we have addressed the first two Principles of Syntax: choosing effective subjects and impactful verbs. In order to address Principles 3 and 4, we need to consider the basic building blocks of sentences—clauses.

A **clause** has a subject and a verb. We can further divide clauses into two types:

- **Independent clauses:** Contain a subject and a verb and forms a complete idea.
- **Subordinate (or dependent) clauses:** Contain a subject and a verb and do not form a complete idea. Begin with a subordinating conjunction (**Example:** because, if, when, which, etc.)

An independent clause contains the most important idea in a sentence, the subordinate clause contains a less important idea.

Understanding the difference between independent and subordinate clauses can improve our grammar and mechanics, but even more importantly, it can help us better convey our ideas. We signal to our reader the relationship between our ideas through the types of clauses we choose. *In other words, our syntax—our sentence structure—conveys our logic—the way we connect our ideas and points.* It creates our logos appeal for the audience.

It can be overwhelming to talk about syntax because there are so many ways to build a sentence. However, a good sentence does not mean a long or complex one. Our audience will judge our writing less on the complexity of our syntax than on our content, diction, tone, and the variety of sentence structures that we use.

In fact, most sentences fall into one of eight types, which I discuss in the remainder of the section.

Eight Sentence Types

- SV.
- SV, and SV.
- SV; SV.
- S V and V.
- S and S V.
- Sub, SV.
- SV, Sub.
- S, Sub, V.

Simple Sentences

Simple Sentences contain only one independent clause—a single idea. A typical paragraph will usually contain at least a couple of simple sentences.

◎ SV ______________________.

Many educators have pointed out the lack of medical evidence to support the existence of learning disabilities and the problematic nature of the "discrepancy method" in labeling a child.

The simple sentence above is relatively long but still easily readable. Notice that the subject and verb come immediately at the beginning of the sentence, as Principle 4 suggests. The sentence contains a long predicate with modifiers, but because these modifiers come at the end of the sentence, it is not confusing.

Compound Sentences

Compound Sentences contain two independent clauses. Because these clauses are both independent, the ideas they contain are of *equal* importance.

◎ SV, and (but/or/yet/so) SV.

This process assumes that underlying biological impairment causes the learning disability, but medical examinations cannot detect such subtle differences in brain function.

◎ SV; SV.

Some individuals with a learning disability may "get by" until they reach college; other individuals with a more severe disability may remain unable to process basic words throughout their lifetime.

These sentences require a punctuation mark to separate the two clauses: a semicolon; or a comma with a conjunction. If you choose to use the semicolon, you should not include a conjunction.

Writers often make one common mistake when they are building a compound sentence: they use the word, "however," as a conjunction. However, the word, "however," is not a conjunction but a transitional word, as I have used it in this sentence. It cannot connect two independent clauses. The word, "but," reflects the same relationship between ideas and can be used to connect independent clauses.

Compound Subjects and Verbs

Compound Subjects and Verbs contain only a single clause, a single idea, because there are not two subjects or verbs.

- S and SV.

  Neurologists and educational professionals have been studying learning difficulties for more than two centuries.

- SV and V.

  Other terms besides learning disability were proposed during the Foundation Phase but did not gain general acceptance.

Notice that these sentences do not contain any commas because they include only a single clause. *Another common mechanical error is to separate a compound verb with a comma.* However, this usage sends a wrong message to the reader—that there are two separate ideas when there are not.

Subordination

Subordinate clauses begin with subordinating conjunctions (after, because, if, when, which, etc.). Subordinate clauses are less important than independent clauses. Therefore, the less important idea goes in the subordinate clause.

- Sub, SV.

  If they have the appropriate documentation, students on campus can be provided with modifications such as extended time on tests or a note-taker.

- SV, Sub.

  The group approved the term, learning disabilities, which rapidly gained acceptance in educational circles.

- S, Sub, V.

  The current term, which was embraced by educators and parents alike, was the product of a grass-roots effort to gain support for struggling students.

We can write many sentences with a subordinate clause multiple ways. For example, we could rewrite the final example above:

- The current term, which was the product of a grass-roots effort to gain support for struggling students, was embraced by educators and parents alike.

Neither version of this sentence is better than the other at the level of style or content. However, the first version emphasizes the idea that the term, learning disabilities, was the product of a grass-roots effort. This version therefore might be the superior choice for a paragraph about grass-roots advocacy in education. The second version emphasizes the idea that the term was embraced by both educators and parents. This version therefore might be the superior choice for a paragraph about coalition building in education.

Rule 3 of the Principles of Syntax is particularly significant when discussing sentences like this final example, which include subordinate clauses between the subject and verb. Sometimes, this type of sentence is necessary, when the subordinate clause modifies the subject. In the above example, the subordinate clause modifies the single word, "term," and must immediately follow it. But these subordinate clauses should remain relatively short: the more words between the subject and verb, the more difficult it will be to read.

Writing Strategy

You can make revision (and grammar) more concrete by listing the eight sentence types on a piece of scrap paper and comparing it to your own sentences. This process can help you to identify improper comma use, run-ons, and fragments without feeling overwhelmed with grammar rules. More importantly, it can help you determine why a sentence is not working and break down ones that are overly wordy.

In addition, the following guidelines can help ensure that your prose is as reader-friendly as possible:

1. *Limit your sentences in most cases to two clauses.* If it has more than two clauses, a sentence is far more likely to be wordy, awkward, or ungrammatical.
2. *Separate clauses with commas.* If it isn't a new clause, it probably doesn't need a comma (with the exception of lists and certain situations where they're optional).
3. *Do not overuse subordinate clauses at the beginning of sentences.* You can move some of these clauses to the end of the sentence, as they will often read clearly in both places. This strategy will also increase your sentence variety.
4. *Limit subordinate clauses at the beginning or middle of sentences to about 5–7 words.* A long subordinate clause at the beginning of a sentence will make it much harder to read.

Exercise

Choose a paragraph from a piece of writing you are currently revising. Analyze its syntax:

1. Determine the structure of each sentence: Which of the eight types is it?
2. Tally the number of each type: how many are simple sentences? Compound?
3. Which sentence structure do you use most frequently? Which do you not include at all?
4. Combine two simple sentences, or change one sentence into a different type.
5. Shift one subordinate clause at the beginning of a sentence to the end.

RHYTHM AND ELEGANCE

An audience often judges the success of a piece of writing on its logos appeal, both at the level of content and style. Paragraphs and sentences need to be built in a way that is orderly and efficient for reading. But as readers, we also recognize elegance in language, are drawn to a piece of writing because we recognize its beauty.

You don't need to be Virginia Woolf to write sentences that will appeal to your audience at the aesthetic level. Elegance in prose requires not an artistic gift but a sense of balance, emphasis, and rhythm. In other words, it is in large part our ear that determines the pathos appeal of a piece of writing.

This section addresses three characteristics of prose that create elegance and appeal to the aesthetic sense of the audience: emphasis; punctuation; and balance.

Emphasis

In a sentence, the primary point of emphasis occurs at the end of the idea. This characteristic of the language is called **end focus:** It stresses certain content and helps build normal sentence rhythm. Consider this short passage by Tom Junod:

> *Once upon a time, a long time ago, a man took off his jacket and put on a sweater. Then he took off his shoes and put on a pair of sneakers. His name was Fred Rogers. He was starting a television program, aimed at children, called* Mister Rogers' Neighborhood. *("Can You Say … Hero?";* http://www.esquire.com/entertainment/tv/interviews/a27134/can-you-say-hero-esq1198/)

The final words of each sentence are impactful and significant: sweater, sneakers, Fred Rogers, and *Mister Rogers' Neighborhood.* More importantly, these words are carefully chosen: Junod's piece is a profile of Mr. Rogers, and no words invoke a more powerful image of the man than a sweater and sneakers. Only at the end of the sentence would these words gain such significance.

You were likely told at some point in your academic career that you should not end a sentence with a proposition like "of" or "to." Like most rules of style, this one has exceptions, but the advice is generally sound. If you conclude with a preposition, you've lost the opportunity to end your sentence with a striking and meaningful word.

You can improve your end focus and manipulate the emphasis within a sentence by adopting four strategies:

1. *Focus the end of the sentence by trimming unnecessary words.*
2. *Shift peripheral phrases* (like "Once upon a time" or "After our visit") *to the beginning of the sentence.*
3. *Shift new ideas to the end of the sentence.*
4. *Avoid repeating the same words at the end of nearby sentences.* A sentence will seem flat if it ends with a word that you used nearby, because your voice will trail off. **Example:** Instead of repeating the noun, use a **pronoun.** The reader will at least hear emphasis on the word just before **it.**

Correctly using end focus not only places emphasis on the most important ideas in a passage but creates rhythm.

Strategies for Shifting Emphasis

Ordinarily, you will want to eliminate empty words from your sentences. However, strong writers often break this rule to change which ideas in a sentence get stressed. This strategy can be extremely effective—but only if you use it sparingly.

When empty words begin a sentence (specifically "it," "there," or "what" expressions), the emphasis changes to the word or phrase immediately following them.

- "It ..."
  - **End focus:** The dog ripped *my favorite scarf.*
  - **Empty words:** It was *my favorite scarf* that the dog ripped.
  - **Empty words:** It was *the dog* that ripped my favorite scarf.
- "There ..."
  - **End focus:** A crocodile is living *in the sewer.*
  - **Empty words:** There is *a crocodile* living in the sewer.
- "What/Where ..."
  - **End focus:** We heard *a colony of bats.*
  - **Empty words:** What *we heard* was colony of bats.

Queen Latifah uses this strategy in a passage from her memoir, *Ladies First:*

> *I am not a psychologist or* sociologist. *I don't have any* degrees, *and I'm not an expert on* life. *What* I am *is a young black woman from the inner city who is* making it, *despite the* odds, *despite the obstacles I've had to face in the* lifetimes that have come my way.

The passage is emphasized through the use of italics, in most cases through the use of end focus. However, Latifah manipulates the third sentence by adding a "what" clause, emphasizing the phrase, "I am." This passage is primarily about her sense of identity, so this choice highlights the main idea of the passage. But more importantly, it also helps create rhythm by varying the sentence structure.

Punctuation

Our ability to manipulate the emphasis in a sentence is perhaps our most important tool in building rhythm—but not our only one.

Punctuation between clauses and phrases can be used to build rhythm and give more (or less) emphasis to the secondary parts of the sentence. In other words, punctuation builds meaning and serves a rhetorical function.

We use three primary types of punctuation to help us build rhythm, each of which sends a different signal to the audience about the emphasis that should be placed on a clause:

- **Dashes:** Clause or phrase has more than ordinary emphasis
- **Commas:** Clause or phrase has ordinary emphasis
- **Parentheses:** Clause or phrase has less than ordinary emphasis

Commas are the foundational mid-sentence punctuation, separating clauses, phrases, and series. Dashes and parentheses, however, can be used as alternatives, giving a passage more variety and greater rhythm.

Dashes and parentheses can be used in two primary ways:

1. *Dashes and parentheses can be used as an alternative to replace commas.*
   - The black-eyed-susan (prolific and hardy) brightens any early summer garden.
   - The rose—beautiful but fickle—is the bane of any gardener.

2. *Dashes and parentheses can be used to connect full sentences in a compound sentence, unlike a comma.*

    - The storm swept across the Midwest—it left fallen trees and flood waters in its wake.
    - The tomato is Nature's greatest gift to mankind (only garlic can claim to be its equal in the kitchen).

The use of punctuation to build rhythm is subtle but significant. Consider the following passage by Gloria Anzaldúa:

> *I will no longer be made to feel ashamed of existing. I will have my voice: Indian, Spanish, white. I will have my serpent's tongue—my woman's voice, my sexual voice, my poet's voice. I will overcome the tradition of silence. ("How to Tame a Wild Tongue")*

The passage begins with a simple, direct sentence with only an ending period. The following two sentences share a similar structure—an independent clause followed by a series of three items. The second sentence uses a colon to introduce the list, which contains single words. The third sentence uses a dash to introduce the list, which includes longer phrases. The passage gains rhythm through this repetition, but it also builds to a climax. As readers, we feel the passage build in intensity, both through the choice of the dash and the longer list. The final sentence, the most significant, brings the essay to a dramatic close. But without her use of punctuation in the second and third sentences, Anzaldúa would have been less effective in her use of climax.

Exercise

Choose a paragraph from a text that you have been reading—preferably not a textbook or similarly "bland" genre.

Rewrite or type the paragraph without any punctuation. Add in your own new punctuation, without looking at the original version. There are many possibilities, so be creative!

How did your new punctuation change the rhythm of the paragraph?

Coordination and Balance

Our use of emphasis and punctuation helps us construct sentences with rhythm; sentences that are pleasing to the ear. But in order to have a fluid rhythm, a sentence must also be symmetrical—balanced.

Coordination (or parallel structure) is the repetition of syntactical structures: strings of nouns; strings of prepositional phrases; strings of predicates; strings of subordinate clauses; a series of sentences with the same subject and sentence structure; or even a series of paragraphs that begin with parallel sentences.

While a sentence must be "parallel" to be grammatical, coordination in a sentence (or parallel structure) is about so much more than grammar:

1. *Coordination helps to emphasize the relationship between ideas:* It makes the equality of different parts of the sentence visible and audible to the reader. In other words, balanced sentence parts look and sound the same—so they share the same level of importance.
2. *Coordination builds rhythm in prose*—that elusive quality, "flow."

Because it is so important in creating rhythm and linking ideas, coordination occurs within sentences, across multiple sentences, and even across multiple paragraphs. There is no limit to the ways in which this syntactical repetition can be used.

At the sentence level, coordination is most commonly used to:

1. *Balance parallel ideas in a series.*

   Example: Attending a hockey game is a unique auditory experience: The rhythmic **whisper** of the blades across the ice; the loud **crack** of the puck against the glass; the harsh **ping** of the puck off the goalpost; and the crowd roaring happily.

In the sentence above, the final item in the list is not parallel with the other items. It has a noun-present participle structure, while the others have an adjective-noun structure. In order to have a pleasing rhythm (and to be grammatical), the final element would have to be changed to, "the happy roaring of the crowd."

2. *Balance parallel ideas presented as pairs with coordinating or correlative conjunctions.*

   Example: The puppy was not only a cuddly ball of fur but also destroying the house.

In the sentence above, the parallel structure revolves around the pair of correlative conjunctions, "not only ... but also." However, the second element of the sentence differs in structure from the first. The first element ("cuddly ball") has an adjective-noun structure, while the second has a present participle-object structure. In order for the sentence to be effective and grammatical, we would have to change the second element to, "a wrecking ball in the house." This revision would make the sentence not only more symmetrical but also more rhythmic, with the repetition of the word, "ball."

Conjunctions

- **Coordinating conjunctions:** and, but, or, nor, for, so, yet
- **Correlative conjunctions:** not only ... but also, either ... or, neither ... nor, both ... and, whether ... or

We can also use coordination and parallel structures in my complicated ways. The following passage appeared in Sarah Adams' short oral essay, "Be Cool to the Pizza Dude" on NPR:

I am the equal of the world,

not because of the car I drive,

the size of the TV I own,

the weight I can bench-press, or

the calculus equations I can solve.

I am the equal to all I meet because of the kindness of my heart.

(http://www.npr.org/2005/05/16/4651531/be-cool-to-the-pizza-dude)

Adams' passage uses two different parallel constructions. First, she builds to a climax through a parallel list of "unimportant" possessions and skills, each of which is longer than the previous one. Second, she echoes the phrase, "I am the equal," in the first and last sentence; this strategy further builds the sense of climax and creates symmetry. This use of balance and coordination is crucial to Adams piece, because its oral form requires not only correct grammar but also a sense of rhythm for the radio.

Writing Strategy

To check for parallel structure, first identify the type of sentence you are looking at. Does it have compound verbs? Is it a compound sentence or a string of independent clauses? To answer this question, identify the main subject(s) and verb(s).

Once you know which type of sentence you are looking at, you can begin to circle any conjunctions and repeated words—those elements around which a sentence will be coordinated. Particularly when a sentence is complex, you should diagram the sentence, listing the series of parallel elements beneath each other.

CONCLUSION

Sentence-level revision often seems imposing. Hopefully, this chapter provided you with strategies for revision, but the process does take time. You will need to build time into your writing process to revise at the sentence-level. I suggest that you revise only a couple of pages at each sitting, reading the paper aloud. Highlight sentences that you stumble over or have to read twice, and revise them after you are finished reading. Before long, you will develop both a process for your revisions and a stronger ear for prose.

It might seem that elegant prose is only a necessity for those who imagine themselves to be writers. But a sense of style is crucial in building our ethos, constructing a logical argument, and pleasing our readers' ear. In other words, style is crucial in writing texts that are audience-friendly, meant to be read by others rather than just express our thoughts. It is our obligation as writers (and you are a writer!) to provide our readers with work that is both efficient and, hopefully, beautiful.

Public Speaking

12 Chapter

As scholars who present at professional conferences in our field and as instructors of classes with a significant public speaking component, we have seen many presentations. And some of the most memorable presentations we have seen over the years begin by using narrative, by telling a story. So in the spirit of those noteworthy openings, this chapter also begins with a story: In this instance, a story about why there is a public speaking chapter in a book about writing and rhetoric in the first place.

To tell that story, one has to go back several years, to around 2007, to the time during which our University as a whole was rethinking its priorities by overhauling undergraduate general education curriculum. At that time, the University of Kentucky general education program had not been reconsidered for several decades, so much of the discussion during this moment centered on what students in our classes would need in order to leave our campus ready to confront the challenges of life, work, and citizenship in the 21st century. The result was the *UK Core,* a substantial portion of which is fulfilled by the Department of Writing, Rhetoric, and Digital Studies Composition and Communication courses, which recognize the interconnectedness of written, spoken, and visual communication.

The logic behind WRD's Composition and Communication course design was to give students the bedrock of skills that they would need to become effective communicators in any modality—be it the written word, a podcast or audio essay, a video, a speech, or even in a medium that does not yet exist. The foundational knowledge set behind all of these modes of communication is *rhetoric.* As you may have read in chapter one, rhetoric has been defined in various ways over the years, from Aristotle's classical definition of rhetoric as the ability to see the available means of persuasion in a situation to Krista Ratcliffe's contemporary notion that rhetoric is "the study of how we use language and how language uses us."[1] However one defines rhetoric, the production of meaning and its relation to language takes center stage. With that being the case, rhetoric in the written word and rhetoric in public speaking have been closely linked. Indeed, for almost its entire history of formal study in the West, rhetoric has concerned itself with public speaking, even to the point of leading a first-century Roman rhetorician Quintilian to say, "Rhetoric is the art of speaking well."[2] Therefore, to give students experience in a broad range of rhetorical situations, the skill of speaking before a live audience becomes integral to the mission of Composition and Communication.

ORGANIZATION

In many, if not most ways, the process for composing a speech closely mirrors the process for writing anything else. The writing process in general is explained in more detail in Chapter 4, "Strengthening Your Writing Process." On the whole, however, a quick overview of the writing process reveals several steps that occur as a person creates: invention, organization, drafting, and revision and editing. The steps in the writing process are often nonlinear, occurring and recurring at different times throughout the life of any project. For example, when I am drafting a speech, new ideas occur to me even as I try to document existing ideas, which means that even though I am ostensibly in the drafting stage of the writing process, I am also in the invention stage, coming up with new ideas as I go. Given this recursive nature of the writing process, you might find yourself revising portions of your speech as you plan an organizational pattern. Even so, our main focus in this section is to give you strategies to organize a presentation effectively.

Speeches differ from most other forms of communication because they are often delivered live, which does not afford the audience the opportunity to rewind or reread a portion that might be difficult to understand. With that being the case, the organization of ideas in any presentation must be easy to follow because it is important that your audience understand each main point the very first time they hear it. To help you with such considerations, it is sometimes beneficial to think of your speech as existing in three distinct sections: A beginning, which gets the audience's attention, introduces your topic, and gives an overview of main points; a middle, which develops those points; and an end, which summarizes the presentation's main ideas and leaves a lasting impression.

Beginnings

"All beginnings are difficult," taught the Rabbinic sage Ishmael ben Elisha. A similar Irish proverb opines, "Making the beginning is one third of the work." American expatriate and poet T. S. Eliot wrote "In my beginning is my end." Though perhaps these maxims are not specifically referring to the opening of a speech, they nevertheless seem relevant, and the general consensus is in: getting a speech started is a challenge. Notwithstanding this difficulty, there are some general qualities that the introductory portions of good speeches often share, and including them in your presentation as a way of structuring your ideas can often help you set your speech off with a bang.

- **Good speeches get our attention.** Sometimes, half of the battle in a presentation is connecting with an audience, and experienced public speakers realize that most audience members decide, consciously or not, within the first minute of a talk whether to tune in or to tune out. The most well-thought-out research and the most compelling argument on a topic will still fall flat if a speaker does not interest their audience right from the start. To accomplish this, many speakers use **attention catchers,** brief, often colorful openings designed to entice an audience into listening. The audience, the occasion, and the topic of a presentation often dictate the kind of attention catcher a speaker will use, but they often range from the humorous—a short joke or anecdote—to the shocking, to the unexpected, to the pressingly relevant. Similarly, many speakers will rely on the eloquence of others, starting their talks with a pertinent quotation. Take a look at the YouTube video below, which contains examples of various attention catchers. What rhetorical appeals are these speakers using to connect with their audience early in the speech? *(https://www.youtube.com/watch?v=2_VvIr1KkLo)*

- **Good speeches often clarify their topic right from the start.** Just like many writers often do not wait until the end of an essay or website to let readers know the main focus of their project, skilled speakers most often present their topic and or argument clearly and succinctly right from the start of their speech. Doing so can help to increase your audience's comprehension by helping them identify a clear area of focus for the talk. Of course, there may be instances in which a speaker wants to work deductively, arriving at a conclusion or thesis after careful consideration of particular instances. However, even in these situations it is wise not to leave the audience guessing at your topic and motives for too long; such uncertainty can result in a talk that seems too unfocused to be effective.
- **Good speeches give an overview of what is to come.** Since the primary audience for many speeches only get to hear a speaker's talk once, it is important that they be able to follow the logic and organization of the talk. Towards that end, experienced speakers frequently included a **preview of main points** in the speech as a way of helping the audience understand the main ideas that the presentation will consist of. Such previews do not have to be elaborate or complex, but including a short sentence or two that highlights the most important ideas in a talk before you develop them will help make the presentation more memorable through repetition.
- **Good speeches build good ethos for the speaker early.** The most effective presentations are ones delivered by speakers an audience feels they can trust. In classical rhetoric, this phenomenon is known as *ethos*—an appeal based on the perceived character and credibility of the rhetor. Many factors influence an audience's impressions of a speaker's ethos—from appearances, to manner of speaking, to seeming familiarity with the subject matter. And while a speaker will most likely never be able to account for every factor affecting her ethos, she can nevertheless get the speech started on the right foot by speaking confidently and showing genuine engagement with her subject matter.

Middles

The middle of a speech, also called the body of a speech, is where most of the speech resides. The content and organization of these middle sections can vary wildly from one speech to another, depending on purpose and occasion. However there are some general guidelines you can apply to most public speaking occasions that will help you organize the body of your speech.

- **Good speeches are organized around main ideas.** When organizing the content you have generated for your speech, it helps to single out the ideas that you believe are most important to conveying your message and achieving your goals. Think of these main ideas as the building blocks of your speech, as something akin to body paragraphs in an essay.
- **Good speeches make effective transitions between main ideas.** If part of a successful speech is presenting your ideas and research in a manner that the audience can follow, some of the most important tools toward that end are **transitions.** The best transitions perform two distinct functions simultaneously: they summarize important information that has just been conveyed, and they provide a preview for the next main point. Consider for example this transition statement, taken from a speech about cultural standards of beauty: "Now that we have spent some time examining the ways that the media constantly bombard us with unrealistic representations of beauty, we can begin to articulate the ways that people's daily lives are negatively affected by these representations." Reading this transition now, without even hearing the rest of the speech, we can clearly understand the main point that preceded it (media representations of beauty) and the main point that will follow (negative effects on people's everyday lives). Providing such summary and preview in between your main points provides the repetition and reinforcement needed to keep your audience engaged and easily able to follow your talk.
- **Good speeches continue to build good ethos in the middle.** Consciously or not, audiences continually keep tabs on a speaker's credibility. Building credibility does not depend solely on any one aspect of the speech, but instead consists of a network of indicators: familiarity with the subject matter, citing authoritative and reliable sources, documenting your logic and reasoning, and showing that you are invested in your topic are all ways to improve your ethos. Being mindful of the way your credibility is established or undermined by your main points and how you develop and support them is crucial to the success of the body of a speech.

Endings

Just as it is crucial to draw an audience into a speech by capturing their attention at a presentation's outset, the way a speaker ends her speech also makes an important impression on an audience. When drafting a conclusion for a speech, here are some ideas to keep in mind.

- **Good speeches often conclude by reiterating important points.** Think of this reiteration as a mirror version of the preview of the main points you shared in the beginning of the speech. Proficient speakers realize that repetition is one tool that helps increase retention, so it stands to reason that summarizing the main ideas of your talk will help emphasize the main take-away points for the audience.
- **Good speeches end memorably.** You have put a lot of time into your talk. You have come up with a topic, conducted research, organized and drafted a quality speech, and you have rehearsed the presentation until your delivery is smooth and confident. Even so, the overall effect of your presentation can be seriously diminished if the speech ends with a whimper. To avoid this all-too-common pitfall, make sure that you craft an ending for your speech that will leave the audience thinking about your topic for the right reasons: that the last few sentences resonated with them, impressing upon them the importance of your topic, rather than a speech that ends with an abruptly with no conclusion at all. The concluding sentences are called **perorations,** and they perform the same function for the end of a speech that an attention catcher does at the start.

Organizing a Group Presentation

All of the above qualities of good presentations are applicable to group speeches. Good group speeches still have a beginning, middle, and an end. They still get and keep their audience's attention and still build good ethos. In contrast however, group speeches present their own set of challenges unique to multi-speaker coordination. Here are some tips to help you organize such a team-delivered talk.

- **Decide on the main points of your speech as a team.** Even though you will eventually assign speaking roles for your group speech, make sure that the preliminary work—the invention and prewriting—is done collaboratively. Making sure each group member has a say in the direction and content of a speech right from the start will ensure a smoother process overall.
- **Organize your speech around main points.** This step is no different from how you would approach a solo presentation. Even so, as you begin to organize around the main talking points, individual speaking roles will begin to suggest themselves. For example, one person might feel most comfortable delivering introductory material,

while another speaker might feel herself best suited for summarizing the background and related research the group has gathered. It is difficult to assign speaking roles before the main beats of the presentation have been worked out.

- **Monroe's Motivated Sequence is your friend.** In the 1930s at Purdue University, a speech teacher named Alan Monroe pioneered an organization pattern for presentations and marketing campaigns that drew upon the rhetoric of persuasion. In this style of organizing a speech, orators walk an audience through a five-stage organization pattern that is seemingly simple yet still proves to be effective. To persuade an audience to take action on an issue, Monroe argued a presentation needed to do five things: garner the audience's attention, establish that a problem exists, propose a solution for that problem, help the audience visualize what that solution might be like, and finally, issue a call to action that informs the audience what they can do about it. Consider the following preliminary outline that uses Monroe's Motivated Sequence on a speech about a single payer healthcare system.

    - **Attention catcher:** What if I told you that our nation could spend considerably less on healthcare while simultaneously enjoying a longer, healthier life?
    - **Establish problem:** Even after the Affordable Care Act, thousands of Americans suffer without adequate healthcare.
    - **Propose solution:** One approach to this issue is what is called a single payer health care system.
    - **Visualize solution:** Consider the success of the single payer health care system as shown by the increased life expectancy and lower healthcare costs in these nations who have adopted the model.
    - **Call to action:** Support legislation and candidates who are in favor of single payer healthcare.

To be sure, the content of this speech is not quite there yet, but after organizing the speech into categorical sections based on Monroe's Motivated Sequence, individual speaking roles have already begun to become apparent.

When organizing a speech using Monroe's Motivated Sequence as a group, it is often a good idea to designate one group member as a **moderator.** A moderator's role during the delivery of a presentation is usually to begin and end the speech, and to introduce each new speaker along the way. In the above example, the moderator could begin the talk by capturing the audience's attention before introducing the group member who will establish the problem. Coming after an attention catcher, such an introduction and transition might look something like this:

> *To better acquaint us with the continuing need for healthcare reform, Hannah will now walk us through some of the problems with healthcare as it now exits.*

Similarly, after Hannah has established the problem, the moderator might sum up her main points and transition to the next speaker and section of the talk:

> *Thanks Hannah. Now that we have seen how healthcare reform is still needed because of continually rising healthcare costs and gaps in insurance plan coverage, Tyra will explain the ways in which a single payer healthcare system might be a solution.*

The group speech moderator often continues on in such a manner, alternating between speakers, summing up and looking ahead, until finally concluding the speech. A well-rehearsed group speech moderator can generate the unity and cohesion that a group presentation really needs to shine.

AUDIENCE ANALYSIS

Chapter two of this book covers audience more extensively than does this chapter, and everything mentioned about audiences and the two-way street of communication in that chapter holds true for speeches you will deliver. Nevertheless, the specific rhetorical situation of delivering live speeches merits a few additional words on audience.

Your WRD class will prepare you for and perhaps give you opportunities to speak to a number of different audiences. However, in most instances, you'll deliver speeches to audiences limited to the classroom environment: namely, the instructor and your fellow classmates. On the surface, that might appear to be all you need to know about the audience for a speech in your WRD class. Think of some of the things you can likely assume regarding this audience: They are college students or faculty at the University of Kentucky, most of them probably live in Lexington or nearby, they have been exposed to the same ideas you have in the class this semester, they most likely have a high school diploma or the equivalent. All of these common factors, taken together, begin to provide us an idea of who we will be talking to, which in turn, clues us in on some of the design choices we can make when crafting a speech—things like what level of vocabulary we can use and what cultural references might be understood or not. However, closer examination of this audience can be enlightening because, even in a classroom that appears to be filled with many commonalities, there can still be vast differences in life experiences: Everything from large differences based on the race, class, or gender of an audience member, to seemingly small but nevertheless consequential things, such as whether or not your neighbor has eaten breakfast that day.

Knowing that the audience for a message can have a powerful influence on the ways a speaker composes and delivers that message, how can rhetors analyze their audience when such groups can vary so vastly as individuals? Speakers in these situations develop their material to be delivered to what is known as a **target audience,** which consists of the individuals the speaker intends to connect with or persuade through a presentation. The notion of the target audience is an often fictitious but nevertheless educated guess at who will be reached through a particular rhetorical exchange. Therefore, when drafting a speech, speakers often use a list of generative questions to help them pin down to whom they will be speaking and which design strategies would be best to reach that audience. Such a list of questions appears on page 33.

Exercise: Considering Audience through Interview

Pair up with a partner in your class and conduct a short interview with him or her regarding a potential topic for a speech. What does this person already know about your topic? What would this person like to know about your topic? Does she or he have any strong opinion on the matter? Why or why not? After noting your interviewee's answers to such questions, you will begin to have a better idea of who the person is as an audience member and how you might best present your topic to such an audience member. Finally, ask yourself if this person's knowledge or experiences with your topic might be typical for other audience members or unique to them.

WORKING WITH VISUALS AND PRESENTATION AIDS

Audience First

When possible, many speakers choose to incorporate visuals and other presentation aids into their presentations. Because you'll be thinking about your audience from the moment you begin designing the presentation, continue keeping their needs in mind as you choose presentation aids and work within your constraints. Every member of your audience will have different needs and abilities, and you may not know just by looking at your audience (or even asking them to complete a survey) what those needs and abilities are. Think about this:

- If an audience member is unable to see or read visual content, will he/she be able to understand my presentation?
- If an audience member cannot hear or understand the auditory content, will he/she be able to understand my presentation?

- Does my presentation require any active participation that makes assumptions about the physical abilities of my audience (e.g., that they can run or walk, that they can eat a snack I bring in, etc.)?

Some of these questions you can answer by polling your classmates or your audience. However, sometimes you can't do this, and sometimes you don't know exactly what questions to ask. For this reason, strive to create presentation aids that will work for all abilities. This is a principle that scholars of teaching and learning call Universal Design.[3] You can incorporate elements of Universal Design by the following:

- Adding captions to video clips or songs
- Making typefaces large and easy to read
- Adjusting images for clarity and sharpness
- Commenting on and discussing images; not just leaving them on the screen
- Offering multiple ways to participate in your presentation

One of the great things about Universal Design is the evidence that it helps all kinds of learners, not just those with physical or learning disabilities. For instance, captioning will allow audience members who are deaf or hard of hearing to understand auditory content used in your presentation, but the captions are also helpful for learners who retain written information more easily than spoken and English language learners. In addition, captions can reinforce for all of the hearing audience the content they're receiving through audio. Everyone wins!

Constraints

Audience needs won't be the only factors you'll need to consider. Like all rhetors in all rhetorical situations, you will have constraints. The length of time you'll be speaking, whether you can use notecards or an outline, what kinds of presentation aids you can use: These constraints will probably be determined by the assignment. These and other constraints, such as the size of your presentation space and the available technology, will shape how you develop presentation aids. Consider the following:

- What technology is available in the room I'm speaking in? Can I use PowerPoint? Prezi? Show video clips?
- How reliable is that technology? Should I have a handout or other backup plan?
- Will I be able to connect to the Internet? What if Internet access is disrupted during my presentation?
- How big is the room? Will I be able to move around?
- Can I use a clicker or a pointer?

Think about your constraints before and during your process of creating the presentation aids, and, equally important, have a back-up plan. It's no fun to spend time creating an elaborate and beautiful presentation in Google slides only to discover that the computer you've got to work with won't connect to the Internet.

Presentation Aids: Media and Design

In today's classroom and professional settings, mention of a presentation often connotes the use of PowerPoint, Prezi, or other digital slideshows. Presenters may embed photos, videos, sound and video clips within their slideshows, but the overall format remains dominant. The ubiquity of PowerPoint and its ilk has given rise to a condition professional speakers, business people and comedians alike call "Death by PowerPoint." Death by PowerPoint occurs when a presenter uses slideshow software ineffectively, creating text-dense, hard to read, redundant, uninformative, and/or aesthetically unpleasing or "busy" slides. The boredom instilled in the audience by these kinds of presentations results in "death": the complete and total loss of the audience's attention.

Watch these videos for two different takes on what constitutes "Death by PowerPoint."

- TED Talk on Death by PowerPoint *(https://www.youtube.com/watch?v=Iwpi1Lm6dFo)*
- Comedian on Death by PowerPoint *(https://www.youtube.com/watch?v=MjcO2ExtHso)*

No one sets out to commit Death by PowerPoint (or Prezi), but it happens. One reason it occurs is that presenters sometimes begin to think of the PowerPoint *as* the presentation, rather than one part of it. This can result in gaffes like reproducing on the slide word for word what you plan to say, reading from the screen, and creating far too many slides. Another reason Death by PowerPoint occurs is the opposite problem: taking the PowerPoint for granted, assuming your speaking is the presentation and the PowerPoint is just what you're required to project behind you. Avoid both of these mistakes. Your presentation is everything you do in front of the audience; presentation aids and speaking are inextricably connected. A few basic principles can help you avoid Death by PowerPoint.

Simplicity

Keeping your presentation aids simple so that your audience can focus on you. Simple PowerPoint slides will attract rather than distract your audience.

- **Keep text to a minimum.** Some presenters suggest no more than six words on a line, no more than six lines per slide. Others reject this so-called 1-6-6 rule in favor of no more than 15 words on a slide or even no more than six words total[4]. There is no hard and fast rule, but less is often more in this case. Stick to key words and main ideas; avoid scripts and multiple subpoints.

- **Many presenters favor sentence fragments,** beginning with active verbs, to avoid too much text. You can see below how Figure 1's text, "Promotes active learning" is more succinct than Figure 2's text. Figure 3's text is even shorter, "Active learning," a key term that you can explore more fully as you're speaking. Whether you use fragments, single words or phrases, or short sentences, be consistent. Continue using the same verb tense and basic structure.

Universal Design
Promotes active learning

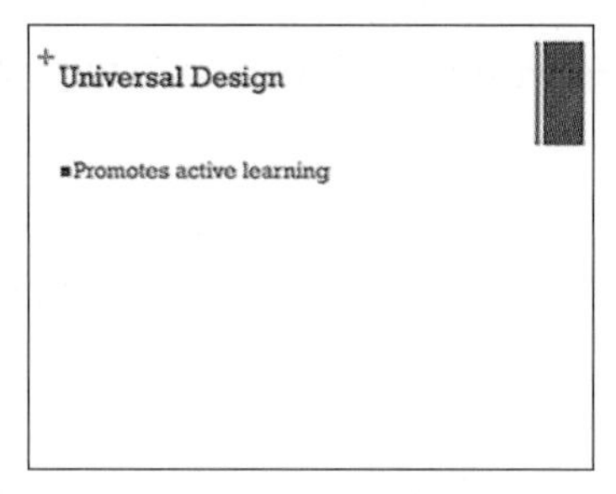

Figure 1.

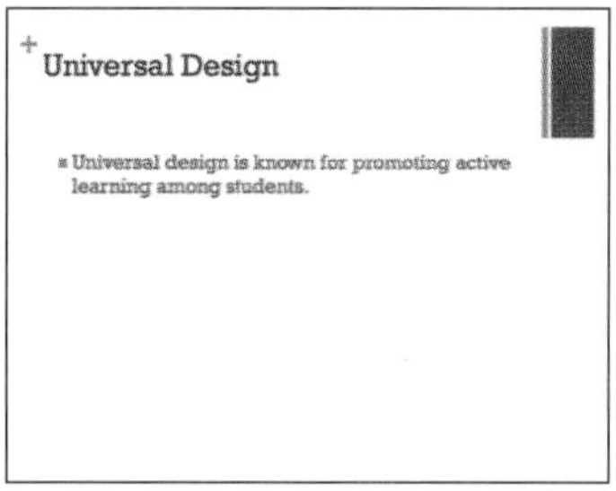

Universal Design
Universal design is known for promoting active learning among students.

Figure 2.

Universal Design
Active learning.

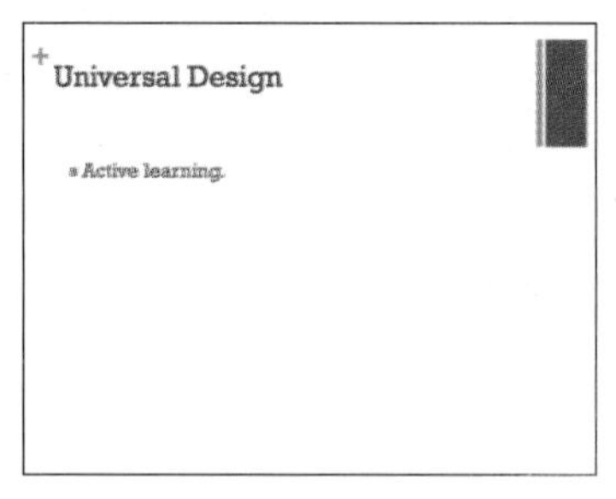

Figure 3.

- **Allow plenty of white space on slides with text.** Consider using blank space, such as a black slide projected behind you, when you want to call attention to what you're saying at that moment.
- **Limit your total number of slides.** Even a well-designed presentation with limited text, lots of interesting images, and appropriate animations will be undermined by too much clicking.

Continuity

As with keeping your presentation simple, maintaining continuity in your design will also help your audience focus on you rather than an ever-changing array of fonts and colors on the screen behind you.

- **Use a single theme or color scheme for your entire presentation.**
- **Limit yourself to two fonts:** One for your titles and one for any text beneath. You may use just one for both, adjusting size and boldness. Choose sans serif fonts that are easier for your audience to read on a screen.
- **Limit animations and transitions.** Using the fanciest fly-ins for text or pictures might backfire if you project on a computer different from the one you designed on; they may appear too slowly or not at all. It can also be distracting.

Variety

It might seem a bit contradictory, but some variety in your slides is a good thing. Vary your use of text and images. Don't click through four slides in a row filled with the maximum amount of text. A text dense (or image dense) slideshow can become monotonous.

- **Use different kinds of visuals:** Video, infographics, charts, pictures. See below about how best to use these.
- **Incorporate other kinds of presentation aids and avenues for participation.** Kristi Hedges of *Forbes* magazine suggests using social media, props, handouts, and group discussion to avoid Death by PowerPoint.[5] You might not be able to have a four-minute group discussion about your topic during your presentation for class, but you can ask a question or two that the audience can respond to briefly.

Visuals

Choosing smart visuals for your presentation is not an easy task. A Google image search for "cancer research" will bring up lots of options, from the very specific and technical diagrams of cancer cells to the stock images of white-coated scientists swirling blue liquids in test tubes and flasks. You might choose the former so that you can explain to your audience how immunotherapy targets cancer cells differently than traditional chemotherapy and radiation. With the proper citation of the image (discussed a bit later), that's a fine use of the image. However, you might also be tempted to use the white-coated scientist, who is probably not a scientist at all, to fill space, to have something on the screen other than text, to meet a visual requirement. This is not a great use of a visual; it's somewhat relevant to your topic, and, if you cite it properly, you haven't done anything unethical by using it. However, stock images like this are usually unoriginal and uninteresting. Even more importantly, stock images are often evidence that presenters haven't thought much about how they want the audience to interact with the image.

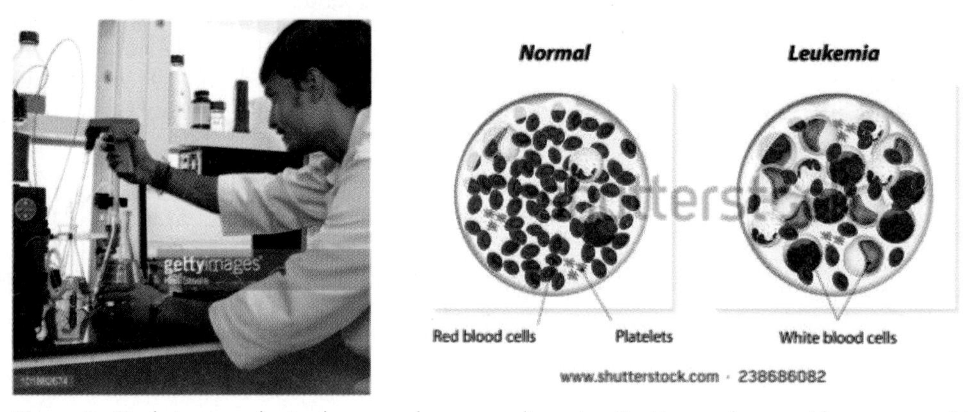

Figure 4. Stock images don't always grab your audience's attention and are seldom memorable. Photo credits: Getty images and Shutterstock.

Remembering that the presentation is *everything* you're doing in front of the audience, you might realize that there's more than one way for audience members to interact with an image. When you pair an image with sound, whether through using a video clip or through your own speaking, you increase the possibilities for interaction even more. Scholars of film and the movies recognize that there can be different relationships between what moviegoers see and hear. Siegfried Kracauer made distinctions between what he called "parallel" and "contrapuntal" sound.[6] Parallel sound matches the image on the screen, such as when an eerie-sounding Theremin plays during a tense moment in a horror movie. Contrapuntal sound offers a counterpoint or contrast to what the image reveals. A famous use of contrapuntal sound occurs in Stanley Kubrick's *A Clockwork Orange* when the protagonist, Alex, happily sings the upbeat "Singin' in the Rain" while attacking a man and a woman in their home. Though Kracauer focuses especially on sound, the terms parallel and contrapuntal refer to the relationship between sound and image more so than just sound. These categories are not the only options for relationships between visuals and sound (see below), but the categories should make you pause and reflect: Not every image is the same, nor do they have to be used in the same way in a presentation.

Going back to our immunotherapy and cancer research topic above, you can imagine a presentation using both parallelism and counterpoint. If you wanted to emphasize the rigors of traditional cancer treatment (e.g., radiation and chemotherapy), you could achieve parallelism by showing an image of a person with cancer who is suffering from some of the side effects, such as hair loss. As you spoke, your audience would see that the effects you describe match the image they are seeing projected. On the other hand, you could project an image of former President Jimmy Carter while discussing the side effects of chemotherapy and radiation. As your audience pays attention to your speech, they may wonder why President Carter is being projected. Because you're using the image as a counterpoint to what you're saying, you can touch briefly on how Jimmy Carter survived brain cancer (at age 90) without enduring chemotherapy, thanks to immunotherapy. The contrapuntal image can be just as memorable and evocative as the parallel image.

Figure 5. A memorable contrapuntal Image. Photo credit: Erik S. Lesser, European Pressphoto Agency.

Informative Visuals

Photographs and videos are of course not the only kind of visual available to you as a presenter. Depending on the context and the topic, many presenters choose to incorporate informative images such as charts, graphs, and diagrams. These kinds of images take us back to the principles of Universal Design: They're meant to convey information in a new way so that more audience members understand the content and understand it better than they would without it.

When creating or choosing an informative image, you want to consider the best format for doing so. Different kinds of figures illustrate different points about data. For instance, if you wanted to illustrate a trend, such as the fluctuations in the average amount of student loan debt carried by undergraduates upon graduation over the past twenty years, a line graph is appropriate. A pie chart, which is used to distinguish the constituent parts that make up a whole, wouldn't illustrate this information effectively. When creating an informative figure, consider the following guidelines:

- Use bar graphs when you want to compare and contrast several different entities that are not part of a whole; e.g., to compare student loan debt for mathematics majors to expected income.
- Use pie charts when you want to explain and represent the different parts that make up a whole. For instance, if you wanted to, you could create a pie chart that illustrates what you spend money on each month. The pie (the whole) is all of the money you spend in a month, and the slices of the pie represent different categories or items of expense.
- Use line graphs to show trends and changes over time. For example, the line graph mentioned above showing the average student loan debt each year for twenty years would illustrate whether that average has increased or decreased.

Other Props

When they're feasible, props can be great alternative presentation aids. If you're giving a presentation on the history of photography, photographs of a stereoscope might be appropriate (and interesting), but bringing in your grandmother's stereoscope and allowing the audience to view images in 3-D through it would be more memorable. Of course, not every presentation lends itself to this type of prop use, and you should always consider your audience's needs and other constraints when bringing in a prop. However, small uses of props can frequently be incorporated. Getting your audience's eyes off the projection screen (and on to you) can increase their attention and engagement with the presentation.

Exercise: Plan B

With a partner, brainstorm back-up strategies for technological and other difficulties you might have on presentation day. Sketch briefly for your partner what presentation aids you're going to use: PowerPoint, videos, props, etc. Then, your partner will suggest one possible mishap that could occur: e.g., campus-wide Internet blackout. Come up with a strategy that will allow you to present in spite of this difficulty. Run through five mishaps and five solutions each.

Exercise: Contrapuntal PowerPoint

Using your outline or speech notes, brainstorm five contrapuntal images or videos you could use in place of images you have now planned. What other changes would you need to make to your presentation to make sure the counterpoint was understood by your audience?

Verbal Citations

Using visuals almost always requires giving credit to a source: a photographer, videographer, or even the authors of a study who provided you with the data for the pie chart you made. However, images are not the only sources you'll be citing in your presentations. Often, you will use summaries, paraphrases, and quotes of outside material to support/develop/explain the points you make in your presentation; you may even be asked to give a research update or present on the research you've done while working on a larger project. Just like when you cite your sources in written work, you need to cite sources in spoken work as well.

Captioning Images

When you use an image created by another person, include a caption on the slide citing the source in the format your instructor requests. In MLA format, images are captioned with parenthetical citations, just like the ones you create for your written work. As a general rule, you should not substitute a URL for a properly formatted caption. A URL isn't very much use to your audience, while a caption with the photographer's last name or the name of the image, might be useful.

Citing Sources

When you reproduce text from a source on a slide, cite it as you would if it were written in an essay. Whether you quote or paraphrase that text, it deserves a citation. Keep in mind, however, that you want to keep text to a minimum.

More likely, you'll be speaking aloud quotes and paraphrases from the sources you use. When you do this, you'll want to introduce and signal the use of your source. Audience members should be able to tell when you are speaking for yourself and when you are quoting or paraphrasing a source. Introducing a source is an opportunity for you to build ethos by introducing the source's author, date of publication, venue, etc., and any details that will show the audience you know your sources and you know why they're relevant. Signaling the use of the source, saying "The authors argue ..." can act as the open quotation marks (or the verbal parentheses for your citation) that let listeners know you're speaking for the source now. Some presenters will say, "Quote" or otherwise overtly indicate a quote is coming before quoting a source. Be wary of doing this. You want to show familiarity with your sources by paraphrasing often and quoting selectively. Plus, there are more imaginative (and subtle) ways to signal the use of a quote.

Finishing Up with Sources

Just like with your written work, you don't want sources to overwhelm your voice. Make sure that after you've introduced, signaled, and quoted/paraphrased a source, you discuss it. Make sure it's clear to your listeners that your voice has returned; you can consider using I-language or taking a brief pause (or both) to signal the end of the quote or paraphrase and beginning of your thoughts.

Once you've done that, though, you're not through with your sources. You should include a Works Cited page (a slide or a handout, or another method your instructor requests) with the citation information for all of your sources, visuals and verbally cited.

Exercise: Yoda Citations

Familiarize yourself with the great character of Yoda, Jedi master of *Star Wars* fame. Use internet research to find out some of Yoda's better known aphorisms. With this list of quotes (and the sources for them), compose a 60-second speech advocating for (or against) Yoda's worldview. In this speech, quote and paraphrase Yoda at least once. When you quote Yoda, do not imitate or mock him. Imagine that your audience takes Yoda very seriously. Perform your speech to a group of 3–4 classmates. Ask them to evaluate how well you introduce and verbally cite your sources.

DELIVERY

When UK basketball coach John Calipari drills his athletes on a play, he does not just have them run through it once. Instead, like all good coaches, Calipari realizes that successful execution of all the elements of a complex play need to be rehearsed, time and time again, in order to refine each element until it has worked its way into a player's muscle memory. The importance of repetition in the delivery of a speech is the same and while the likelihood of you being body checked during a presentation is significantly lower than it would be on a basketball court, without repeated rehearsal of individual elements in the delivery of your speech, you might experience the emotional equivalent of one.

The embodied delivery of a speech can often be one of the last things we think about. Between coming up with a topic, researching, planning, and drafting a presentation, the way that we will look and sound during a talk can easily take a back seat to what seem to be more pressing concerns. However, classical rhetors thought about delivery quite a bit. Indeed, Roman philosopher and political theorist Cicero believed delivery to be such an important part of the success of a message that he singled it out as one of the five canons of rhetoric in his book *Rhetorica ad Herennium.* To become an expert on all the varied nuances of the craft of delivery might take a lifetime, but there are nevertheless several important aspects of delivery that you should be mindful of as you rehearse your talk.

- **Consider the space and your body.** Much of the labor of composing a speech happens from the shoulders up, so it is easy to forget about the other 90% of your body. For this reason, it becomes important to rehearse delivering your speech exactly as you plan to in front of the audience. What will the room for your speech look like? Will you be standing or sitting? Will there be a podium or lectern in the room? How will you incorporate visual aids? Are there planned gestures you wish to make at particular times to emphasize points? Will you need to use speaking notes? All of these are considerations that might affect how you hold yourself during a speech, and recreating these environmental constraints as you rehearse will help you make the most of your rehearsal time.

- **Plan to be nervous.** During the pressure of a live speaking engagement, many (if not most) speakers experience some degree of performance jitters. Such nervousness is OK and is perfectly natural. However, the well-prepared public speaker will be mindful of these nerves and will not allow them to manifest themselves in a nervous tick. As you rehearse your speech, practice planting your feet firmly and speaking without shifting your weight from foot to foot. Similarly, as you prepare for the talk, practice what you will do with your hands to fight the urge to fidget with clothing, notecards, etc. Have a plan for how you will position yourself in the speaking space, and then practice, practice, practice that plan.
- **Make eye contact.** Professional public speakers know that one of the ways they can connect with an audience and build rapport is to talk *to* them, not *at* or *around* them. One of the ways to convey that you are speaking not to a general audience but to individuals within the audience is to practice making eye contact. A goal to shoot for when rehearsing eye contact is to strive to look 90% of your audience in the eyes during 90% of your speech. Obviously, with larger audiences or in some spaces, such a lofty goal might be unachievable. Nevertheless, as you practice your delivery, try to imagine yourself looking at the people to whom you are speaking.
- **You have good ideas; make sure they hear them.** Air conditioners. Jackhammers. People in the hallway. Flatulent trucks. Audience members packing up their backpacks early. That guy eating a cheese Danish out of a crinkly cellophane wrapper. When we deliver a speech, there is often background noise that seems unavoidable. If you have a particularly deep voice or are a soft-spoken person, all the great points you worked so hard to develop might be at risk of getting lost in a sea of static. Because of that, one important aspect of delivery you can rehearse is **vocal projection.** Practice speaking from your diaphragm, not the top of your lungs, and envision yourself projecting your voice to the very back row of the audience. Relatedly, many speakers consciously rehearse speaking at a slow, moderate pace as they practice because of the tendency to speak more quickly under the pressure of a live performance. Rehearsing projection and a slow delivery can go a long way towards ensuring your message gets through to the audience.

The benefit of these rehearsal sessions is difficult to overstate: They give a speaker the chance to practice all of the above aspects of delivery, but they also increase familiarity with the subject matter, which in turn lowers speech day anxiety, boosts confidence, and bolsters credibility. Good public speaking is rarely easy, but setting aside time to practice your delivery can at least make things easier.

PRACTICING AND PREPARATION

A music teacher of mine once said: "Practice doesn't make perfect; *perfect* practice makes perfect." This might seem like a cynical piece of advice if you're struggling to overcome pre-presentation nerves while finishing up your Prezi, formatting your Works Cited, and working on your assignments from your other classes as well. Who can be "perfect" in these circumstances, especially just on a rehearsal day? Nonetheless, there's truth to it. Practicing your presentation is essential, and any amount of practice will pay off. However, practicing without the ability to correct any of your bad presentation habits will not help you pull off a (close to) perfect presentation when the time comes. To practice perfectly, you need to practice prepared, practice with an audience, and practice with specific goals in mind.

Your practice sessions will have greater impact on your performance the closer your presentation is to completion. Should your instructor set aside class time for rehearsal, there are several ways he or she might approach it: Some classes might focus just on introductions or conclusions, while others perhaps give you time to practice individual elements of delivery, such as making eye contact or projecting your voice. Generally speaking however, you should come to an in-class rehearsal prepared as though you had to give the presentation for a grade that day. This means a thorough outline (or speaking notes) and a completed PowerPoint or other presentation aids. If you come to this session with an incomplete presentation, you might be able to get some good feedback from your classmates and instructor about some parts of your presentation, and you may be able to ad lib the missing pieces. However, this rehearsal will not really prepare you for the experience of giving the presentation in full. If you haven't rehearsed it, you might not realize that you don't know how to pronounce the name of a source you decide to add at the last minute; or that your audience doesn't find your opening anecdote as amusing as you had anticipated; or that the transition between your second and third points is a bit awkward; or that you run over (or under) on time when you present the whole thing. Having a complete presentation for each rehearsal can help you find these problems early.

An audience for your rehearsals will also help you improve your performance. A small audience (ideally of your classmates or other people who will be there when you give the actual presentation) can help you overcome jitters by giving you experience on a smaller scale. Rehearsal audiences are also invaluable tools for learning what the audience finds most interesting and engaging. If your rehearsal audience smiles or laughs when you want them to, makes consistent eye contact, or otherwise uses backchannels (nonverbal cues, such as head nods or concerned expressions) to indicate they're paying attention, you know that your presentation is engaging them. On the other hand, if they seem confused or inattentive, you can ask them after your rehearsal if there was anything they didn't understand or had questions about. If you didn't notice much about your audience's reaction, you need to keep rehearsing; you've either not made eye contact with them consistently enough to notice them, or you're still too nervous about/unfamiliar with your presentation to focus on the audience. If an audience isn't available, record yourself presenting with a stable camera (on a phone you can prop up or a webcam; don't attempt to take one long selfie during a rehearsal). If you don't have a camera, rehearse in front of a mirror. Deliver the presentation just as you would for an audience; half-hearted rehearsals don't constitute "perfect" practice.

Finally, make the most of your rehearsals by setting specific goals. Professional public speakers offer a range of "ideal" numbers of rehearsals for perfecting a presentation, with some advising six and others over fifteen. There probably isn't a magic number that will make all speakers equally prepared to present in all contexts. Nonetheless, more practice is probably better than less practice, so try to rehearse your presentation a minimum of five times. Set aside blocks of time that will allow you to rehearse the entire presentation multiple times at a stretch. Begin each rehearsal with a completed presentation (you can rehearse parts of your presentation before the whole thing is completed, but those don't count towards the goal of five) and an audience (or a camera or mirror). Your presentation aids should be ready to go. Have a mechanism in place to time your rehearsal; if your presentation doesn't meet the time constraints, don't count that rehearsal towards your goal of five. Additionally, if you make any major changes to content, set your rehearsal counter back to "1."

Below are some goals you can set for each rehearsal. Add your own goals based on your individual strengths and weaknesses.

- **Rehearsal #1:** Rehearse the entire presentation without stopping.
- **Rehearsal #2:** Make eye contact with each member of your audience at least twice during the presentation.
- **Rehearsal #3:** Be able to continue speaking while transitioning slides, notecards, pages, etc. (Try to memorize what's at the bottom and top of your note cards/pages; make sure you know exactly the order of your slides.).
- **Rehearsal #4:** Play the "Babble game."[7] Instead of speaking actual words, say "Blah blah" or some equivalent (e.g., "Yadda yadda" or "Rah rah") throughout the entirety of your speech. The goal is to use your tone of voice, gestures, and other nonverbal communication to emphasize your points. It should help you think about which words and phrases you want to emphasize and how you want to do it.
- **Rehearsal #5:** Dress rehearsal. Try to replicate the atmosphere of your actual presentation as closely as possible. Wear what you plan to wear that day. This might help you anticipate minor issues, like the difficulty of making eye contact if you're wearing a baseball cap. Try to rehearse in a similar location at a time similar to when you'll actually present. You'll get a good feeling for how the environment might affect you.

Finally, establish a cut-off point for rehearsals and spend your energy in other kinds of preparation. Try to get a good night's rest before and make sure to eat well the day of the presentation. Make your handout copies in advance. Have a backup of your PowerPoint on a jump drive (and/or email it to your instructor). Do something small to make yourself feel confident and relaxed: practice Power Poses, get to class a little early, talk to your parents or a friend about how much you've practiced and how smoothly you're anticipating it will go.

Visualize yourself finished with the presentation and feeling awesome. Think about what you'll do afterwards: get a fancy coffee, Skype with your younger siblings, reward yourself with Netflix. You're ready, and your presentation will be finished sooner than you think.

Watch this video to learn about Power Poses: *(https://www.ted.com/talks/amy_cuddy_your_body_language_shapes_who_you_are?language=en)*

SELF AND PEER EVALUATIONS

When you receive written work back from your instructor with comments, what do you do? Oftentimes, students will read over their own work again as they digest their teachers' feedback. It's a bit more difficult to do this with your speeches and presentations. However, with preparation, you can be a better self- (and peer-) evaluator of your verbal work.

Self-Evaluation

Part of self-evaluation will take place before your presentation; when you're rehearsing and setting goals. You'll have a sense when you finish presenting whether or not you hit those goals. However, you shouldn't take your initial emotions when you finish presenting as the sole indicator of your performance. Sometimes, we're way too hard on ourselves (and sometimes we're way too easy on ourselves). If possible, try to have your presentation recorded. Your instructor may be able to arrange this for you, but if not, you can ask the instructor if it's okay for you to record (video or audio) your presentation.

Before you watch the recording, jot down some notes about your performance. Try to do this as soon as possible after you've presented, even if you don't plan to watch or listen to your recording right away. Make note of what you think went well and what didn't work out the way you wanted. Only after you've done this writing should you view the recording. The initial notes you take can be useful. They can tell you if your experience of the presentation was wildly different from the reality your audience experienced, but they can also help you watch the recording with a grain of salt. Remember that the video (or audio) is a mediated representation of the presentation, not the real thing. How you felt it went does in fact matter.

Note discrepancies between your initial feelings and what you observe in the recording. Did you feel like you were making eye contact frequently, but the tape suggests you turned towards the projection screen … a lot? Do you have trouble hearing yourself? Are you speaking too quickly? Does it seem like you're reading from your notes rather than speaking? It's easy to note these issues when you view a video of your presentation. You can also prioritize working on these issues for your next presentation. For instance, if you have trouble making eye contact, add another rehearsal dedicated to working on eye contact. You can also deprioritize things that went really well. If you're great at eye contact, but speaking way too fast and monotonously, focus on the Babble Game rather than eye contact in your rehearsal.

Peer Evaluation

Whether you're asked to give feedback to your peers or not, preparing to give comments to your classmates is a useful task. It can help you self-assess, as when you notice a peer pausing too long between slides and remind yourself to memorize transitions. It can also help you cultivate good listening skills; a really underrated, but vitally important talent. Chances are good that you'll end up working with some of your classmates in group projects during the semester. Listening to their presentations can tell you a lot about their interests, knowledge, and skills.

When listening to your peers' presentations, the first thing to remember is to respect them. If you've already presented, you know how difficult it can be to be at the front of the classroom, trying to impress your class and instructor, and trying not to do anything to embarrass yourself. You may have noticed a classmate who looked bored or inattentive, maybe even one who was checking his or her phone or looking at flashcards for another class. Perhaps you understood that this person's inattention probably had nothing to do with the quality of your presentation. When you're giving a presentation, though, even the most neutral look of boredom on an audience member's face can be devastating. Try to emulate the behaviors of an attentive audience member.

- *Attentive audiences arrive on time and stay seated during the presentation.* If you're late (or need to leave early) wait for the applause following a presentation to enter (or leave).
- *Attentive audiences silence and put away phones and tablets.* Clear your desk of electronic devices unless you need them to follow the presentation.
- *Attentive audiences make eye contact with the presenter.* Remember that the presenter is trying to make eye contact; it's a two-way street.

- *Attentive audiences participate in the presentation.* When you're asked to respond, respond. Take notes so that you can formulate questions to ask during the Q&A (if applicable) or offer feedback (if you're watching a rehearsal).

Your instructor (or your peers) may ask you for more substantial feedback during rehearsals or presentations. If so, remember that you shouldn't focus exclusively on the presenter's delivery. You may notice a lack of eye contact or numerous vocalized pauses ("uhhs" and "ums"), but don't let that stop you from learning something from the presenter. Your instructor may give you specific criteria for evaluating your peers. If not, consider the following questions in formulating your response:

Feedback on Content

- What did you learn from this presentation?
- What would you like to learn more about following this presentation?
- Which sources did you feel contributed most to the presentation?
- What visuals did you find most memorable, appropriate, or interesting?

Feedback on Organization and Structure

- How did the presenter begin the presentation? Was it engaging and memorable?
- Could you list the major points of the presentation? Did the presenter list these at any point in the presentation?
- How did the presenter conclude the presentation? Could you tell when the presentation was ending? Were the final words/sentences memorable?

Feedback on Delivery

- Did the presenter make eye contact with the audience?
- Were the presentation aids visible and aesthetically pleasing?
- Could you hear and understand the presenter's voice?
- Could the presenter do anything to make the presentation more accessible for all people?

You might not have time to answer all of these questions, or to talk to the presenter about all of them. If you have the opportunity, tell the presenter something important he/she did well and one or two important things he/she can work on. You can also work in your interest in the presentation topic, but make sure you prioritize helping the presenter. Remember that an audience is an invaluable tool for presenters looking to improve their performance.

NOTES

[1] Ratcliffe, Krista. "The Current State of Composition Scholar/Teachers: Is Rhetoric Gone or Just Hiding Out?" *Enculturation* 5.1 (2003). Web 29 February 2016.

[2] Quintilian. qtd in "American Rhetoric: Definitions of Rhetoric." *American Rhetoric: Definitions of Rhetoric.* Web. 18 March 2016.

[3] Sheryl Burgstahler, "Universal Design: Process, Principles, and Applications." *Disability, Opportunities, Internetworking, and Technology.* University of Washington. N.d. Web 11 March 2016.

[4] John Zimmer, "PowerPoint Math: The 1-6-6 Rule." *Manner of Speaking* http://mannerofspeaking.org/2010/03/04/powerpoint-math-the-1-6-6-rule/

[5] Kathy Hedges, "Six Ways to Avoid Death by PowerPoint." *Forbes.* N.p. 14 November 2014. Web. 11 March 2016.

[6] Siegfried Kracauer, *Theory of Film: The Redemption of Physical Reality.* New York: Oxford UP, 1960.

[7] Nick Morgan, "Seven Ways to Rehearse a Speech." *Public Words.* 26 July 2012. Web. 11 March 2016. <http://publicwords.com/2012/07/26/seven-ways-to-rehearse-a-speech/>

Chapter 13

Visual Rhetoric and Design

> *We are all designers. We manipulate the environment, the better to serve our needs. We select what items to own, which to have around us. We build, buy, arrange, and restructure: all this is a form of design.*
>
> —Donald Norman, from "Epilogue: We Are All Designers," *Emotional Design*[1]

This chapter focuses on visual literacy, visual design, and visual rhetoric. **Visual literacy** is the ability to observe and interpret visual information. This draws on our skills of visual observation, our abilities to notice and recall small and subtle visual details. We use our visual literacy to interpret the storyline of a film, or to laugh at the humor in a comic. When we use a website, we use visual literacy to infer what is a button that can be clicked, and we infer from the button's color and shape what that click might do or lead to. **Visual design** is the practice or profession of using design elements, like light, color, or shape, to make those films, comics, or websites into something we can interpret. **Visual rhetoric** combines visual literacy and design with rhetoric to make a branch of rhetorical studies concerned with the persuasive use of images, whether on their own or in the company of words.

In this chapter, we will practice and sharpen our observational skills. We will look closely at visual artifacts made by artists, photographers, filmmakers, author-illustrators, advertisers, and designers; and we'll talk about how writing and images can work together to form a narrative or argument.

We'll also show you tools and skills that will help you design your own visuals from scratch. Feel like you are not a visual designer? *Au contraire!* We all have some hand in the design of our own homes, rooms, or appearance. You may not sew your own clothes, but you make choices about what clothes you wear.

In 1917, Marcel Duchamp submitted a urinal, signed "R Mutt," as a work of art to be exhibited at the Society of Independent Artists in New York. It was not accepted (but also not officially rejected—it was dumped behind exhibition screens where it could not be viewed), and instead shown at the studio of his friend, Alfred Stieglitz. Duchamp called the urinal and other works like it, including a bottle rack and a bicycle wheel on a barstool, "readymades."[2] Duchamp did not make any of these sculptures. Instead, he claimed that art does not stem from what is made by the hands, but through the choices made in the mind of the artist:

> *... you can cut off the artist's hands and still end up with something that is a product of the artist's choice since, on the whole, when an artist paints using a palette he is choosing the colours. So choice is the crucial factor in a work of art. Paintings, colours, forms, even ideas are an expression of the artist's choice.*[3]

Masterful painting skills or an expert understanding of computer animation code aren't prerequisites to being an artist or a designer. We all have the potential to make art, to design, or to create visual stories or arguments. The potential lies in the choices we make. This chapter will guide you through those choices, how to interpret those choices made by another, and how to make your own choices when creating a visual artifact.

Why did you wear what you are wearing today? You may have chosen something chic and color-coordinated, because aesthetics are important to you. Maybe you're wearing shorts with several pockets because it's warm out and you have a lot of gadgets to keep on hand. If so, then functionality is important to you. You might be dressed in business wear for an interview, or in your favorite team's jersey for game day. If so, then you are conscious of an audience around you—you chose your outfit to send a message.

If you're not sure of a strong purpose behind your choice of clothes, consider this: "Why *aren't* you wearing something else?" Why gray running shoes and not yellow flip-flops? Why a t-shirt with a lion's head instead of plain black? Why an oxford shirt and slacks and not a bathrobe and spandex leggings? Why not go to class naked? Are these choices made for aesthetic reasons, practical functional reasons, or due to audience consideration?

> Not every visual object is visual rhetoric. What turns a visual object into a communicative artifact—a symbol that communicates and can be studied as rhetoric—is the presence of three characteristics. ... The image must be symbolic, involve human intervention, and be presented to an audience for the purpose of communicating with that audience.[5]

Using the three characteristics in the quote above as a guide, do you think there are parts of your wardrobe that constitute visual rhetoric? Why or why not? Similar questions can be used to examine and interpret visual artifacts, and to help us design our own. Why use a drawing instead of a photo? Why use bright color instead of black and white? Our first answers might be *"because it looks better,"* but often, that's not the reason. As with well-planned works of writing, well-planned visuals are created with a purpose in mind, and the choices made in the design of the visual build up to achieving this purpose.

CLOSE LOOKING

If we practice close reading to understand and interpret how a text communicates a message, then to understand and interpret how a visual communicates a message, we practice close looking. The readings and activities in this section will help you sharpen your observational skills.

Exercise: Sherlock Holmes

Clothes are a part of appearance that a person can choose. Even if choices are limited by authority or available resources, we may personalize, adapt, or search for a better option. Whether intended or not, clothes give our audience (everyone around us) clues about occupation, age, regional identity, and even aspects of personality.

View or **read** the opening scene from "The Adventure of the Blue Carbuncle" in *The Adventures of Sherlock Holmes*. The famous detective makes detailed description of an unknown person based on a hat. Some of his conclusions are based in the era of the times. Which conclusions could still apply today? Which seem too far-fetched or highly assumptive?

Holmes uses steps of logical reasoning to draw conclusions about the person's life and habits. His mind combines an observation, of things seen on the hat, with a generalization, something learned from life experience, and then reaches a conclusion. This is an example of inductive reasoning. (Note: We're used to hearing about Sherlock Holmes' powers of deductive reasoning, but the techniques the character employs are often inductive … maybe the "Inductive Detective" didn't have that snappy alliterative quality Doyle wanted.) Here's a few of Holmes's "deductions" broken down:

Exercise: Sherlock Holmes (continued)

Table 1. A table of Holmes's "deductions," showing how he made them from observations.

Observation	Generalization	Conclusion
There are grizzled hairs stuck to the inside of the hat.	When wearing a hat, one's loose hairs come off inside.	The wearer of this hat has grizzled hair.
It has a flat brim with curled edges.	These hats were all over london a few years ago.	This hat is a few years old.
The hat material is of good quality.	Hats of this quality cost money.	The wearer had money.
The hat is battered and dirty.	Most people would throw out a hat this worn.	The wearer has no other hat, perhaps cannot afford it.
The hat is very large.	Large headed people are smart.	The man is intellectual.

The catch here is in the generalization. Generalizations may come from life experience, or may be told to us by someone else. If we use a faulty generalization, our conclusion will be wrong. In the last row of the table above, Holmes observes that the hat is large, and draws the conclusion that the man is intelligent. The television adaptation plays this moment as a conversational joke, but phrenology, a theory that claimed to determine personality characteristics from the shape of the skull, was a debated science throughout the 1800s.

All of us use steps of logical reasoning. (Did you catch that generalization?) Our minds usually move through the process quickly, so we don't always note each step of making an observation, combining it with a generalization, and coming to some conclusion. Breaking the process down into steps can help us see what visual details we observe and what past experiences might have led to true or false conclusions.

Exercise: Sherlock Holmes (continued)

Try your hand at playing Sherlock Holmes, and see how much you can guess about someone based on an object. Select an object you use often and have with you: a shoe, watch, hat, backpack, keychain, pair of glasses, etc. Whatever the object is, you should be comfortable having a stranger examine it inside and out. Avoid objects like notebooks or journals where someone can simply open and read all about you. If the object is a cell phone, lock the phone so no one can see your contacts or messages. Trade your object with another person or group of people.

After you've traded objects with someone else, make a table on a piece of paper with three columns labeled "observations," "generalizations," and "conclusions." Look at the object given to you, and start listing all the different traits the object has in the observations column. No trait is too simple to matter, so start with the obvious. Note what the object is, and the object's size, color, texture, shape, and weight. Look closer. What condition is it in? How old is it? Does the object have a specific physical function? What is it, and can it perform it well? Are there other objects attached or inside? Is there a particularly well-worn area of the object? Is the object right-handed or left-handed? What does the object smell like? Is there writing on it? Has anyone altered this object to change it from its original design? Is there anything about this object that has a specific cultural association (for example, a sports logo, cartoon character, brand name, election slogan, etc.)?

Clothes are a part of appearance that a person can choose. Even if choices are limited by authority or available resources, we may personalize, adapt, or search for a better option. Whether intended or not, clothes give our audience (everyone around us) clues about occupation, age, regional identity, and even aspects of personality.

Figure 1. An illustration of Holmes examining the hat in "The Blue Carbuncle."

Exercise: Sherlock Holmes (continued)

After you've looked as close as you possibly can, make conclusions based on each observation. If you can, provide the generalization that led you to this deduction. (You may find yourself thinking of the conclusion before the generalization!) Allow yourself freedom to go out on a limb, to write down what may seem like a far-fetched idea. Try to fill up the entire table.

Now for the spine tingling part! Share your conclusions with the class. After sharing, ask the owner to reveal which deductions are correct and which are incorrect. Make note of the incorrect deductions.

Discussion and Questions

What assumptions or false generalizations may have lead to incorrect deductions? In our example, the Holmes' deduced that the owner did not also own a laptop or smartphone. They assumed that a person wouldn't own two or three items that perform the same task. They correctly deduced that the owner used math in class but was not a math major, and that the owner has a busy lifestyle and is a little careless. Many of the Holmes' deductions seem plausible but aren't correct because they assumed the object had not been owned by anyone else in the past.

- While playing Sherlock Holmes, what conclusions did you make that were incorrect, and where did the logical reasoning process go wrong?
- What incorrect conclusions were drawn about you when your own object was examined? How do you think the person playing Holmes came to this conclusion?
- What observable aspects of your object did you, the owner, intentionally choose or design? Why?
- Write about or talk about a time in your life when either (a) you made a false assumption based on someone's appearance or (b) someone made a false assumption about you based on your appearance.
- We've talked a lot about how false generalizations can get us into trouble. When might observations, generalizations, and conclusions help us out in a positive way?
- How can we regulate our own logical reasoning processes so they stay useful? How do we keep it from causing bias or discrimination?

LEARNING TO SEE

The same close-looking skills used in the Sherlock Holmes activity can also be applied to visuals and works of art. These skills are important enough that one of the largest police forces is training its officers to see with more skill. Through a special program at the Metropolitan Museum of Art, Amy Herman teaches a course called "The Art of Perception." New York police officers observe and discuss artworks from the museum to practice observation skills and communication skills. Let's try our hands at "the art of perception" and describe what we see.

Figure 2. Photograph by Mark Steinmetz.

Exercise: What's Going on Here?

Take a look through **the following gallery of images by the photographer Mark Steinmetz** *(http://www.hartmanfineart.net/artist/mark-steinmetz).* To boost your observation skills, try looking at the image and drawing it on a piece of paper. The drawing doesn't have to look anything like the original for this to help. This process will slow down your eyes and cause you to focus on specific areas of the image. With each picture answer and discuss: What is going on here? What time of day is it? Who is this person, and what is his/her life like? With each question and answer, be precise about what exact part of the image gives you a clue. Make notes on your drawing, if you wish. Look closer. Are your conclusions correct? Are there alternative conclusions that can be drawn from your observations?

... and How to Talk about It

Observing is just one part of visual literacy. To really understand how visuals communicate, we need to know how they carry a message. The message might be in its subject matter or content: what the picture is of—such as a dog—or what the subject matter might symbolize—such as loyalty. This message will differ based on the image's context, its surroundings: a cute puppy on a grade-schooler's notebook has one purpose, a cute puppy on a poster at a rescue shelter has another. Media may change the message: A photograph says something different than a finger painting. The message can also be molded by formal elements like light, color, shape, and texture. A dog in bright light gives the viewer a different feeling than a dog in darkness.

CONTENT AND CONTEXT

Content is the subject matter of an image, or what the image is about. No image is without content. Even an abstract painting of one large blue square has content: that of a large blue square.

Context refers to the varied circumstances in which a work of art is created and interpreted. Think of context as anything happening outside the boundaries and surface of the image. The painter's life, the materials available, the time and place in which it was made, the history and culture of that place and time, and the potential audience of the painting, who may have even hired the painter to make it.

Context includes all of those aspects of the image's making, as well as those same aspects of the image's viewing and interpretation. When I step in front of a painting from 1660 and give my interpretation, I become part of the painting's context, outside the image's boundary looking in. My life, the time and place in which I view it, the history and culture of that place and time, and even the setting in which the image sits, such as an online thumbnail gallery or a large museum hall, all become part of the painting's interaction with me, its viewer.

When looking at **Claude Lorrain's Sermon on the Mount** *(http://collections.frick.org/media/view/Objects/220/3516?t%3Astate%3Aflow=ab13ff70-a50a-420a-8a30-b26858602979),* one NYPD officer who participated in "The Art of Perception" said the image resembles that of a potential suicide victim, a "jumper."[6] In reality, this painting is of Jesus Christ on a high plateau preaching a sermon. How did this happen? The painting's context has changed. Its viewer, or audience, is now a police officer. Every day his job puts him in situations that many other people see few times, if ever, in their lives. He may have been called to save potential suicide victims, or had extensive training in this area. Perhaps current events that happened recently in his own local sphere put this topic on his mind. Maybe the fact that he saw the painting as part of a police training exercise, and not while visiting the museum on the weekend, changed what he saw in the expressions of the faces in the crowd watching Jesus.

In 1656, Claude Lorrain definitely did not intend the police officer's vision of a suicidal jumper to be part of the painting's content. This is through no fault of his own because of the great shift in context. He couldn't foresee a New York City police officer viewing this image in a museum in the twenty-first century anymore than he could imagine New York City existing. Lorrain most likely did intend to create a dramatic scene of an individual drawing a crowd through uncommon acts and speech. Perhaps the body language and facial expressions of onlookers seemed similar to the officer. Or, maybe the visual traits Lorrain used to create this dramatic scene, like a dark, backlit landscape, somehow touched a chord in the police officer's mind.

This leads us to a third crucial part of any image or artwork: its **form.**

Exercise: Photo Pictionary, Round One

In the game Pictionary®, players guess a word or phrase from pictures drawn by a teammate. In Photo Pictionary, players guess a word or phrase from a photograph. Each team will create a photo, using a smartphone or digital camera.

First, form into teams. Next, randomly select a word. For this game, avoid simple words and names of objects that are easily found in your area. Uncommon action words, adjectives, or more abstract nouns work great: "panic," "hero," "revenge," or "apathy." Try this **game word generator** *(https://www.thegamegal.com/word-generator/),* set to hard or really hard, for inspiration.

Teams get ten minutes to take a walk, discuss a plan, and shoot a photo that illustrates their word. Players can pose, use props, or photograph something in the environment. Text of the word is, of course, not allowed. Only one photo can be used for guessing.

Once photos have been taken, teams trade photos with another team. Players on each team discuss possible guesses for what word the photo is illustrating, then submit to the class their best guess. Take one or two attempts at guessing the word (don't reveal the answers just yet!), then move on to Photo Pictionary, Round Two.

FORM

Form can be difficult to talk about, because it is so simple. Formal elements of art and design usually sound like something we learned from *Sesame Street*: shape, color, light and dark, size, texture, to name a few. We can also think of form as what's left when we ignore the subject matter. Form is all the parts of an image that are not of semantic significance. Form has nothing to do with visual symbols that we translate to meanings with our minds, and everything to do with sensations we perceive physically with our eyes and bodies. Varying the formal elements of a visual can completely reframe or even overturn the content of the image.

These images portray similar subject matter: a Doberman Pinscher looking at the camera. There are a few content differences: The ears are in different positions. To viewers who are familiar with dog body language, or with the practice of cropping Dobermans' ears, this content may mean something. Is the dog alert? Angry?

Figure 3.

Figure 4.

Now, let's look at differences in form. Which picture is darker? Darkness can trigger a feeling of the unknown, or even potential danger. Which picture has more color? Color can make something more inviting, or add warmth or coolness. What textures do you see in the picture? How might they contribute to the viewer's interpretation?

One of these pictures is taken from a different point of view. Point of view is a particular aspect of the design element of space: the real or illusional space and distance in the image. How does this point of view change what message might be in this image of a dog?

Reading: Formal Elements of Visual Design

To include form in our toolbox of visual literacy, we need a good grasp of the different formal elements of visual design. Most of us have had reading and writing in school. We were taught the basic forms of language: letters, sounds, words, meanings, sentences, tenses, etc. We were taught that words have a meaning (content), and that the meaning can be flexible or changed, that the words may have different connotations to different people or connotations that change over time (context).

We also learn that writers might choose a word for an aesthetic, sensory-based purpose: to make a rhyme, or a harsh clacking sound, or a smooth swooshing sound, when writing poetry. The formal elements of visual art are like poetry, they play on our most basic reactions to visual stimulation. Darkness is mysterious, we can't see objects in it. Large objects have power, especially when they are bigger than us. The natural world is curvy, the man-made world is angular and square.

These tutorials *(https://ukwrite.wordpress.com/tutorials/visual-literacy/)* cover the formal elements of visual design. Examples from a wide range of media are included, such as photographs, film stills, paintings, sculptures, maps, diagrams, and architecture. (We'll cover things like unity, emphasis, and grids later on, so skip those for now.)

Exercise: Photo Pictionary, Round Two

In round two of Photo Pictionary, teams can use a photo-editing program to alter design elements, like lighting and color, to better illustrate subtle connotations of the word. For example, if a team's word is "explore" and the guesses were "lost" and "stranded, the team might alter their photo to add some more *positive* connotation.

This can be a fun time to play with Photoshop®, but online photo editors like PicMonkey or Pixlr work great, too. Just be aware of pre-made filters that change several formal elements at once. Make sure that every change is one that nudges the guessers into the right direction.

Exposure, brightness, and contrast will adjust lights and darks. If the word has a dangerous or mysterious connotation, darkening the image might help. Color balance, white balance, hue, and saturation can adjust color or color intensity. If the word is playful, bright colors might help. Rotating and cropping will change the direction of major lines. For example, if the word describes dynamic movement, a slight rotation tilting the horizon and edges of objects on a diagonal might help. If the word describes something massive and overwhelming, a quick collage job that changes the scale of things will help. Adding text is not allowed!

After editing the photos, try a new round of guessing. Did the changes in design elements give the guessers the hints they needed?

Form & Content Activities

Exercise: Whose Font Is This?

The tutorial **"Communicating with Shape"** explores the world of typography. Typefaces come in a wide variety of shapes: some curvy and organic, some blocky and mechanical. One of the tasks of a graphic designer is to create or choose typefaces that subtly send the right message or set the right tone.

Compare the typefaces of brands that are selling similar products, such as two Internet search sites or two soft drinks. How do their typefaces differ from each other? What image or identity do you think the company is trying to project with this typeface?

Exercise: The Formal Side of Humor (NSFW!)

The tutorial "Communicating with Shape" also contains a link to *McSweeny's Inter Tendency. McSweeny's* uses a typeface that connotes serious, studious thinking, and yet, essays like **"It's Decorative Gourd Season, Motherf%$#^rs!"** are anything but serious and studious. It's as if the content of the writing is wearing the visual design, or the "clothing," of another kind of content. Does this "clothing" make the article funnier than it would be **if it were in another font** *(http://www.mcsweeneys.net/articles/im-comic-sans-asshole)?* When a message is wearing an *unexpected visual look*, the result could be very inappropriate, or, depending on the context, very humorous.

Satirical television shows, parodic websites, and other humorous multimedia outlets perform this kind of visual design dress-up to combine content with form in unexpected ways. The ***Onion*** *(http://www.theonion.com)* wears the visual design of a mass-marketed newspapers. The ***Colbert Report*** *(http://thecolbertreport.cc.com/videos/3vijkd/the-word---color-bind)* uses all the visual elements of cable news shows, and in particular, The ***O'Reilly Factor*** **from Fox News.**

Kick back and take a look at some of your own favorite multimedia satires and parodies. What is the content in the piece? What choices have been made about formal elements, like lighting, color, shape, texture, scale, or point of view? How does dressing this content in this form affect the rhetoric of the piece? What cues, visual or verbal, let us in on the joke?

After studying the visual rhetoric of your favorite spoofs, satires, and parodies, try making one of your own. What spoken or written message can you dress in new visual "clothes" for a wildly different effect?

Exercise: Close Looking at Content, Context, and Form

In this section, we've learned to observe visual details closely, to check our observations and deductions for false generalities, and we've explored the three aspects of any visual artifact: form, content, and context. Let's practice what we've learned so far.

Find a visual artifact that holds your interest for several minutes. Look for visual artifacts:

- at online visual archives such as the Library of Congress or the Kentucky Digital Library;
- around your own environment in the forms of posters, flyers, book covers, and magazine articles;
- in your own community in billboards, sculptures, or murals;
- in the media in the form of sets, stills, and shots from film, video, and TV (choose a still or very short segment to work from).

Do a close looking of this artifact. To gather ideas and observations, use the table below as a guide.

<table>
<tr><th>Content</th><th>Context</th><th>Form and Media</th></tr>
<tr><td>what the image is of</td><td>origin, year made, for what purpose</td><td>line, shape, color, light and dark, texture</td></tr>
<tr><td>what is going on</td><td>maker, artist, his or her life</td><td>space or point of view used</td></tr>
<tr><td>setting of image</td><td>where image is kept, exhibited, viewed</td><td rowspan="2">size and scale: of object within, or size relationship between viewer and the work</td></tr>
<tr><td>what objects in the image symbolize or signify</td><td>cultural associations</td></tr>
<tr><td>image title or text in the image</td><td>connotations brought by the viewer</td><td>media: photo, painting, sculpture, billboard, etc.</td></tr>
</table>

Exercise: Close Looking at Content, Context, and Form (continued)

Start by simply looking and writing down your observations. What is going on in the image?

Research the image's maker, origin, and purpose. What was the context when the image was created? What might the purpose have been? What is the new context that you bring as a viewer? Look at the visual artifact's form. What is its color, shape, size, texture, point of view, etc.? Very few visuals use *every* formal element; which seem to be most important to the visual artifact you chose? What effect does the form have on the viewer? Does this effect reframe the content in any special way?

VISUAL ARRANGEMENT

Writers take great care to put their words and ideas in orders that make sense, or create the best effect. A story or argument can be told in organized parts: a beginning, middle, and end; a situation, a problem, a solution. We need unity to have a sense of order, a sense that all the things being talked about are tied together for a reason.

However, there can be such a thing as too much unity. For one thing, too much repetition and order can have a choking, stagnating effect. Perhaps our story or argument is about a conflict or a problem, and unified harmony just doesn't fit the bill; or, maybe we want to spotlight a person, place, or idea and make it more visible. In these cases, we can put the techniques used to create unity in reverse gear and create emphasis instead.

In forming ideas and communicating them to others, we slide back and forth on a spectrum from unity to emphasis: from *"this is very much like that, we are all one people, we can learn from these similar situations"* to *"this is different from that, a new situation has changed this person's life, this situation is urgent because of x, y, and z."*

Visuals also communicate by using unity and emphasis. With repetition and orderly arrangement of visual parts, we get a sense of harmony and unity. In an image where unity prevails, we see the whole before the parts, and our eyes travel the surface of the image as we look. By contrast, when one visual part is brighter, bigger, or shoved off to the side of a group, it is emphasized and we fixate on it. In an image where emphasis prevails, we see one focal point first. Our eyes will travel to other parts of the image, but they are always drawn back to that attention-grabbing focal point.

Figure 5. Andrei Rublev's "Trinity" is an image in which unity prevails.[7] (Tretyakov Gallery, Moscow, Russia/ Bridgeman Images.)

Unity and emphasis can be used within one artifact of communication, or used to arrange and curate massive collections of communications: salon walls, web galleries, magazine spreads, and poster displays. Unity gives balance and order to a menagerie of different works, and emphasis draws us in to notice one work before the others. In this section, we will spend extra time on using unity and emphasis to arrange collections of text and images on the page or screen. This practice is known as **page layout** or **document design.**

UNITY

Within any narrative or argument, we need unity to have a sense of order, a sense that all the things being talked about are tied together for a reason, that they have something in common with each other.

You've probably already worked with unity in your writing. Placing words in close **proximity** to each other in a sentence, as in *"I thought of age, loneliness, and change,"* prods the reader to consider those things as a unified group, even though they don't necessarily always appear hand in hand. Themes, images, and sounds can unify writing through their **repetition:** echoing phrases such as "I have a dream," a story bookended with similar scenes, or sounds that rhyme or have alliteration. Characters or ideas can be neatly connected and **aligned** with each other by simply sitting on the same line of text, in a well-crafted transition sentence or line of poetry. Events that take place at the same time are collected into **containers:** sections, chapters, or between pauses of a speech.

There are techniques of visual unity that relate to these: **proximity, repetition, alignment,** and **containment.** These techniques arrange visual bits into families or groups, and create a sense of visual harmony. **Read the tutorials** *(http://multimediasandbox.org/tutorials/visual-arrangement/)* to learn about proximity, alignment, repetition, and containment and how they work as visual cues.

Exercise: Website or Printed-Page Markup

Check out the **tutorial on unity in document design** *(https://ukwrite.files.wordpress.com/2012/01/vislit-unity.pdf).* Here, proximity, repetition, alignment, and containment are used to order a webpage that holds many different kinds of information. Take a look at a website or print publication that holds dense amounts of information in one page or screen. Mainstream newspapers, magazines, and their companion websites are great examples. Draw a sketch of the page or screen layout, with wiggly lines for text and cartoon squares for images. Note and include any bold type, dividing lines, gutters (empty space), or containment boxes.

After you've sketched the layout, now, mark where proximity, repetition, alignment, and containment are used to let the reader know:

- What text goes with what images?
- What stories or images are related?
- What parts perform similar functions?

It all looks and sounds so simple, doesn't it? However, editors do make tragic mistakes in document design. Check out some of these layout mistakes. How did unity work against the editors' intentions?

EMPHASIS

Emphasis can be thought of as the antithesis of unity. Unity groups visual bits into a unified whole. With emphasis, one part sticks out boldly and becomes a focal point. Our attention is transfixed to that point, everything else takes on a secondary role.

Perhaps you've noticed how emphasis works in written texts. If repetition and similarity create harmony and unity, then a striking **contrast** against that harmonious backdrop will draw our attention: an event of violence in a peaceful setting, a short chapter in a novel full of long chapters, a drop in volume in an otherwise rowdy and loud speech. The **placement** of different parts of writing can also draw the reader's attention. Opening and closing paragraphs are great places to put those scenes or ideas that will stick in the reader's memory. (In visual design, the placement with the most emphasis is near the center, or in a spot that is "pointed to" by other lines and edges.) If proximity encourages us to see things as a group, then moving a word away from others draws attention to it. In this example, the ellipses are very intentional, they **isolate** the word:

> *'Alone' ... The word is life endured and known.*

Isolation, placement, and **contrast** are also ways that we can create visual emphasis, or "focal points." **Read the following tutorial** *(http://multimediasandbox.org/tutorials/visual-arrangement/)* to learn more about these techniques, and how they work in visual narratives and arguments.

Figure 6. Leonardo da Vinci's *Last Supper* employs emphasis: Christ is in the center, isolated from other figures, shapes point towards him, and a bright window creates value contrast. (Santa Maria della Grazie, Milan, Italy/Bridgeman Images.)

Exercise: Poster Scavenger Hunt

Check out **the tutorial on emphasis in document design** *(https://ukwrite.files.wordpress.com/2012/01/vislit-emphasis-1.pdf),* then take a break from the classroom and stretch your legs! The tutorial illustrates how print advertisements of the 50s and 60s used emphasis to draw the reader's attention away from other content. Your campus has several places that are hot spots for posters and flyers, and these crowded poster walls function in a similar way.

Divide into teams, with each team heading to a different poster-populated area. When you arrive at your destination, step far back from the poster wall, or walk casually through the hallway. What poster are your eyes first drawn to? What flyer would grab your attention even if your mind was busy looking for an open study carrel? Which technique(s) of emphasis does the poster use to make it stand out from the other posters? Take a picture, make a photocopy, or take the poster itself (just put it back where you found it when the activity is over) and bring it back to class and share.

MORE ARRANGEMENT TECHNIQUES

While talking about unity and emphasis in writing and visual communication, we also have focused on unity and emphasis in document design. Here are a few more strategies that will help you arrange text and images in a way that best suits your purpose.

Juxtaposition

Juxtaposition is the act or instance of placing two things near each other, side by side, or one after the other. When two things are juxtaposed, our senses pick up on more differences and similarities. Juxtaposing characters in TV and movies emphasizes the personality and characteristics of both: Penny and Sheldon in *The Big Bang Theory*, Liz Lemon and Jack Donaghy in *30 Rock*, or Leslie Knope and Ron Swanson in *Parks and Recreation*. The characters aren't pitted against each other, but each character's traits become more vivid in the other's presence.

Figure 7. These side-by-side images from Mark Laita's photographic work "Created Equal" juxtapose two similarly sized photos of people in similar poses. We can't help but compare and contrast the two. (Photo by Mark Laita courtesy of Fahey/Klein Gallery.)

Figure 8. These images from Nicolai Howalt's photographic work "141 boxers" juxtapose two similar photos of a boxer before and after a fight.

In visual art and design, juxtaposition happens when two things of similar size and type are placed near each other, usually one on the right and one on the left. When the two images are aligned and of the same size and shape, we can't help but see the frames as reflecting mirrors. We scour the image and see differences and similarities we may not have noticed otherwise. Juxtaposition is the driving force behind before-and-after photos, and humorous side-by-side images of celebrities who were "separated at birth."

If you're working on a document that closely examines the similarities and differences of two people, places, events, or ideas, consider juxtaposing two images by cropping them to the same size and placing them side-by-side. Giant multi-sided mirrors of juxtaposition can be created using grids.

Figure 9. Unlike Laita's work, which is shown in side-by-side pairs, Howalt's photographs are displayed in a massive grid, juxtaposing not just before-and-after boxers, but boxers young and old, victorious and defeated.

Figure 10. The grid of Roger Shimomura's "24 People for Whom I Have Been Mistaken" juxtaposes images of different Asian Americans. The viewer is left wondering what the artist looks like, and how he could be mistaken for any of these very different looking people.

Grids

Grids with units of equal sizes create super multi-sided mirrors of juxtaposition, prompting the viewer to compare and contrast among a multitude of items. But the grid is also flexible. Images and blocks of text can vary greatly in size and in importance, and the grid will still organize them into a unified piece.

This works because the grid uses all the key ingredients of unity—proximity, alignment, repetition, and containment—to create ordered arrangements. If you need to arrange a large number of very different visual items, like many fabrics in a quilt or many images in a document, the grid is your go-to resource.

For this reason, grids are the backbone of two (and sometimes three!) dimensional multimedia compositions. **Check out these tutorials** *(http://multimediasandbox.org/tutorials/visual-arrangement/* and *http:/ukwrite.files.wordpress.com/2012/01/docdes-fvh.pdf),* to see how grids (and riffs off the grid) are used to order information in multiple media, and how you can use grids to design a multimedia document. Whether your project is text heavy or image heavy, serious and informational or humorous and opinionated, the grid is an arrangement tool flexible enough to fit any need.

Two Kinds of Balance

Symmetrical balance occurs when the visual elements on two sides of a central axis mirror each other. For example, if you draw a vertical line down the middle of a perfect five-pointed star, each side is a mirror reflection of the other. It is symmetrical. "Perfect" is exactly what visual symmetry is: a geometric ideal.

Figure 11. The perfect symmetry of the Grand Trianon at Versailles in Paris, France.[8]

Asymmetrical balance creates a feeling of dynamic movement, or of natural wildness. But a wild appearance does not mean "unplanned"; both the symmetrical and asymmetrical gardens seen here were skillfully planned by the designer.

Figure 12. Just as the symmetry of the Grand Trianon is perfectly crafted, so is the asymmetry of Prospect Park in Brooklyn, New York.[9]

Asymmetrical balance works by balancing one visually interesting form with another. These two visual bits do not have to be mirror images, or even similarly shaped. *They just need to draw the same amount of visual interest.*

Here's an example: In the top image of Marilyn Monroe, her hair has been parted to create sultry blond bangs on the left, asymmetrically balancing her famous mole on the right. Both of these features were visually arresting—they are in every iconic image of the actress—but they are of opposite sides of her face. The bottom two images approximate what Marilyn would look like with truly symmetrical features.

Figure 13. This image shows Marilyn Monroe's carefully-crafted, signature asymmetrical look (top), as well as her face PhotoShopped in perfect symmetry (bottom).[10]

Check out more examples of balance in the following tutorials, including the use of symmetrical and asymmetrical balance in document design. Your page layout grid can be filled in ways that are symmetrical, or asymmetrical. If you are working on a document that strives for balance and order, try a two-column or four-column grid: it will lead to symmetrical balance. If you are working on a document that cries out for a feeling of movement or drama, try a three-column grid: it will lead to asymmetrical balance.

Visual Hierarchy

Westerners like to think we read left to right, top to bottom, but we learned from the tutorials on emphasis that we are more inclined to look at what is most different from the rest: what is unusually large, brightly colored, in motion, or being pointed at. If you are still in doubt, turn your web browser's ad blocker off and see how much time passes before your eyes are thrown off course by a pink hula-hooping bear touting low prices for pharmaceuticals. That web ad took over the high ground in the screen's **visual hierarchy,** or, its set of focal points that vary in amounts of emphasis.

Visual hierarchy can be very useful. Imagine you are looking for a specific quote that you *know* is in a 600-page printed textbook, but you can't find it in the index. You scan pages page by reading the largest text first, and when we've found the right section, we can bother with the small text.

In the 1800s, large broadsides played games with visual hierarchy to create intentional miscommunications. Large fonts, when seen from a distance, would tell a different story from the real text.

Figure 14. This Civil-War era broadside attracted viewers by posing as a news headline. When the smaller text is read, one realizes it is an ad for quack medicine. This broadside advertises the opening of Indian Territory to homesteaders. Is the odd use of visual hierarchy is an intentional part of its rhetoric?

Any visual medium can have a sequence of visual hierarchy, guiding a viewer from the most attention-getting focal point to more and more subtle focal points, leading the viewer along a nonlinear path, picking out bits of a narrative or communication along the way.

Figure 15. "Harlem" (1946) by Jacob Lawrence shows a city block. The viewer's eyes move from building to building, seeing both comic and tragic events. However, the large white building of the upper class pulls for our attention more than the events in the lives of the less-visible residents.[11] (© 2015 The Jacob and Gwendolyn Knight Lawrence Foundation, Seattle / Artists Rights Society (ARS), New York.)

You can use visual hierarchy in your own document by using different degrees of emphasis throughout the pages. **Check out this tutorial to see more** *(https://ukwrite.files.wordpress.com/2012/01/vislit-emphasis-1.pdf).*

Exercise: Analyzing Page Layouts (starring Bill Cunningham)

Bill Cunningham photographs fashion in New York City: from the runways and high society parties to the sidewalks of Brooklyn and the Bronx. He himself doesn't dress fashionably (he rides a bicycle and wears a blue plastic poncho repaired with duct tape) but his fashion insights and photography are world-renowned.

Every Sunday, in the *New York Times*, Bill Cunningham creates a carefully laid out page of photographs with a central theme. Just as we arrange our clothes for the day to suit our personalities, Bill Cunningham arranges his layouts to suit the topic of his photo essay.

Each of these examples uses a different type of layout. Choose one, and analyze how the layout works. Is it symmetrical or asymmetrical? (One of these uses a type of symmetry we have not included in this text!) How do your eyes move along the page? Which directions do the people walk, face, or look? What are the main focal points? The secondary focal points? What is repeated? What comparisons and contrasts are made?

Read the text in the page. What is the photo essay about? How does the layout suit this topic?

If you'd like to get to know Cunningham better, you can **watch the documentary *Bill Cunningham New York*** *(https://zeitgeistfilms.com/billcunninghamnewyork/)*, check out **his YouTube channel** *(https://www.youtube.com/playlist?list=PL074352F6ACE39E76)*, or look at his slideshows at the *New York Times* website. But to find these page layouts, do a Google search.

Exercise: Your Own Curation

A great way to practice organization of your own multimedia compositions is to play with organizing the writings and photographs of others.

If you've been working on a writing project, you no doubt have a "shoebox full" (or bookmark folder, or Pinterest board) of artifacts relating to your topic. If not, pick something you're a bit obsessed with, or pick something from the list of random topics below. Collect articles, quotes, interviews, photographs, posters, charts, maps, and other materials relating to your topic.

Now, curate and arrange your materials into an orderly layout. You can cut and paste into a word processing document, or if your eyes are burning from screen fatigue, go old school—print it out and use scissors and tape. This is just for practice, so there's no need for polished finished products.

What will be the main focal point of your collection? What will be the secondary focal points? What images and text go together? You can make the reader associate them by aligning them into columns of equal width, pushing them closer together, or putting them in a box. Does a symmetrically balanced layout suit this topic or does asymmetrical balance suit it better? Do you want blank space to give the layout a buoyant feel, or should you fill it with dense amounts of text and information? What about a grid layout that is tilted at an angle? When you are finished, share your curation with the class.

Random Topics to Choose from if You Cannot Think of One Yourself

- Penguins
- Conan O'Brien
- Beyoncé
- Tap dancing
- NASCAR
- Tasmanian devils
- Pool (swimming or billiards)
- Wind power
- Palm oil
- Australia
- Migraines
- Soccer
- Bricks
- All-terrain vehicles
- Honey bees
- Dinosaurs
- NASA
- Astigmatisms
- Mountain biking
- The Westminster Dog Show

Exercise: Photo Scavenger Hunt

Get some photography practice by going on a photo scavenger hunt. Choose a category of subjects that you can find in your community, such as images of horses, things that are blue, drawings of faces, things that make annoying noises (or you may choose a more serious and academic topic, such as product labels that use demeaning stereotypes). Hit the town and grab as many images as you can. Write about 400 words of text about your scavenger hunt results. Arrange your images into a page layout. You can consider the same questions above as you arrange your layout.

NOTES

1 Donald Norman, from "Epilogue: We Are All Designers," *Emotional Design*, last modified March 20, 2003, http://www.jnd.org/dn.mss/CH-Epilog.pdf.

2 "Fountain (Duchamp)," *Wikipedia*, last modified June 9, 2014, http://en.wikipedia.org/wiki/ Fountain_(Duchamp).

3 "An Interview with Marcel Duchamp," *Art Newspaper*, last modified March 29, 2013, http://www.theartnewspaper.com/articles/An-interview-with-Marcel-Duchamp/29278.

4 "'What was the proudest moment of your life?' 'The first time I put on this uniform,'" *Humans of New York*, last modified July 4, 2013, http://www.humansofnewyork.com/post/54627852998/what-was-the-proudest-moment-of-your-life-the.

5 Kenneth L. Smith, Sandra Moriarty, Keith Kenney, and Gretchen Barbatsis, eds., *Handbook of Visual Communication: Theory, Methods, and Media* (New York: Routledge, 2004).

6 Neil Hirschfield, "Teaching Cops to See," *Smithsonian*, October 2009, http://www. catalystranchmeetings.com/art-work/downloads/Smithsonian_magazine_Oct_2009-Teaching_Cops_to_See.pdf.

7 "Trinity (Andrei Rublev)," *Wikipedia,* June 30, 2014, http://en.wikipedia.org/wiki/Trinity_ (Andrei_Rublev).

8 "Grand Trianon," *Wikipedia,* last modified June 11, 2014, http://en.wikipedia.org/wiki/ Grand_Trianon.

9 "Prospect Park (Brooklyn)," *Wikipedia,* last modified June 27, 2014, http://en.wikipedia.org/wiki/Prospect_Park_(Brooklyn).

10 David M. Harrison, "A Study of Asymmetry of Faces," *Upscale,* accessed June 29, 2014, http://www.upscale.utoronto.ca/GeneralInterest/Harrison/Parity/FaceStudy/FaceStudy.html.

11 Jacob Lawrence, "Harlem," *The Jacob and Gwen Knight Lawrence Virtual Resource Center,* accessed June 27, 2014, http://www.jacobandgwenlawrence.org/artandlife04.html.

Chapter 14

How to Make Your Own Documentary

> *Every narrative film is in the process of turning into a documentary, just as every documentary is struggling to become a piece of fiction.*
>
> —Jean-Luc Godard

Let's say you've been tasked with making a short documentary, perhaps as an assignment for your WRD class. (If so, your instructor will determine the specifics of the assignment, and this chapter should be regarded less as template for those exact needs and more of a general overview.)

Congratulations. You're about to learn a little about a time-honored and more effective than ever way to forge an argument—and maybe a lot about yourself. Don't be daunted by the technology. Your equipment can be as simple as your iPad or your telephone. The real raw materials are your imagination, your reasoning, and the unlimited narrative potential of the real world around you.

You're also lucky to be living in a time when the documentary form is thriving and justly celebrated. Films like *Man on Wire* and *Searching for Sugarman* are finding large enthusiastic audiences who appreciate the particular power of the actual stories they tell. And the Do It Yourself aesthetic which has always driven this form is now enhanced by the availability of inexpensive and relatively impressive ways of filming and editing.

ONE: PRE-PRODUCTION

So: What kind of a documentary do you want to make? Will it present a political argument, or attempt to expose a particular social or economic issue? Will it be an autobiography of sorts, an expression of the motivations and tensions which helped make you who you are? You might try to document a particular process, showing us not only how it happens but in what sequence, with what sense of cause and effect, and what the outcomes and consequences are. A cinéma vérité approach calls for as little intrusion as possible—no incidental music, voiceover, graphics, the closest you can get to a fly on the wall. Or would you rather intrude, insert yourself in the center of the story?

Maybe you'd like to use photographs, film, and other archival footage out of context to make ironic connections to the material you've filmed yourself. Or maybe the mockumentary form appeals to you—a piece of fiction, sometimes carefully crafted, sometimes largely improvised, which purports to document a phenomenon which never actually happened.

These are just a few of the almost infinite possibilities. One thing I've learned, after years of teaching narrative film history and now immersion in the alternate universe of documentaries, is that there are if anything even more possibilities with the documentary form than the fictional one. Each subject finds its own form.

Perhaps it is easier at this juncture to think about what a documentary *isn't.* It isn't a news report, for instance, which insists, in its "fair and balanced" way, to present just the facts and only the facts. Unlike traditional notions of journalism, a documentary ought to have a point of view. On the other hand, this doesn't call for reckless bias either. The challenge of balancing logos and ethos here is the delicate art of acknowledging contrary claims and counter-arguments while not falling into the kind of bland, chit for chit argument of "one for you, and one for me" which refuses to take any position and ends up convincing no one of anything.

You don't want to preach to the already converted, but it is futile to worry too much about die-hard partisans on the other side. These issues are much more obvious in political arguments, but even something as tangibly subjective as your autobiography needs to consider such audience concerns as tone, rhythm, and emphasis.

But a documentary is not an advertisement either—a glittering and manipulative attempt to overwhelm and seduce the viewer, a momentarily distracting shiny object to make her crave something she never even knew she needed.

Nor is a documentary a simple report, a summary of facts and figures.

So: Thinking about what your subject is and how you are going to present it is step one. Brainstorm this problem the same way you would any writing project. Here's a good way to get your feet wet, get to know your colleagues a little better, and make yourself a little more comfortable with the process on a very micro-level.

Exercise: The Proust Questionaire

Break into pairs. (This exercise will work best if you choose to work with a classmate you don't know very well.) Now locate a famous set of thought provoking questions called the Proust Questionnaire at the following web link: *http://www.vanityfair.com/magazine/2000/01/proust-questionnaire.* Note that these questions, unlike the usual bland getting-to-know-you formulas, are actually quite probing, resist generic responses, and call for maturity, reflection, and self-awareness.

Your responses are not framed, in other words, to impress each other, your teacher, or most importantly yourself, but to dig down deep, just as most of you are undergoing a huge transition in the coming of age experience. Take plenty of time—think of this as the pre-interview—just talking over your reactions and getting comfortable with your "subjects." Some of the questions may be beyond you—you probably don't know at your age when you were "most happy"—and some of them may just not click. Not to worry. All we really need for this exercise are a good seven or eight answers.

Now you are ready to film the interviews. Set up two chairs so that the subject is looking at the interviewer, with the camera at somewhat of an angle. Let your subject take her time, and try not to accept short, glib, or superficial answers. Give yourself some dense and passionate material to work with. In terms of ratio, it would be nice to get ten or fifteen minutes of material which you will then edit down to around a minute. Once you have a rough cut of the content (all of these steps of the process will be explored in more detail later in this chapter, so read the whole piece first so you can try to apply these ideas from the get-go), go back and drop in some appropriate images, music, etc. to augment the portrait.

Instructors, this process could take a day or two in the opening weeks of the course, but you should set aside another day for the students to screen their work together and begin the assembling of a creative community which can effectively but diplomatically workshop the process.

So now that you've give it a try on a small scale, let's start thinking again about pre-production on your larger project, which you might envision as running five to eight minutes or so. If that doesn't sound like much time to you, consider the issues of editing, compression of material, and even aural density—a minute of well made film is usually the product of hours of careful work.

Think of it this way: Your problem here is almost exactly the opposite of the college student's traditional dilemma of "padding out" a paper. There, the unfortunate result is running on and on for pages extra to reach some minimum. Here, the easiest thing in the world to do is turn a camera on a blank wall for half an hour, but that doesn't make you Andy Warhol. You want the maximum of engagement, intensity, and edification for each moment of your film without resorting to hyper-caffeinated editing or otherwise bombarding the viewer with Too Much Information at once.

As with a chapter, though, try to narrow your thesis as much as possible. Some of the same rules apply: Don't take on too large and unmanageable a topic, find one aspect of a larger idea which interests you most, see what's been done in a similar vein, etc. The more you can focus your topic, the more you can exclude from the net you're casting, the better off you'll be.

Once you've chosen your subject, pre-production can begin in earnest.

Casting is as important here as in any fictional film. You want not just your primary subject but witnesses who will be supporting/refuting/contextualizing that subject to be as vivid, pointed, and intriguing as possible. Although the word "casting" sounds antithetical to what you're doing, just going out on the street and interviewing random observers is more likely to produce a mess than a masterpiece.

Bear in mind that some people just want to be on camera so much that they will be impossible to pull into any kind of serious conversation. Such "naturals" should probably be avoided. On the converse, some very thoughtful and incisive people who could be of great value to your project are going to be shy and intimidated by the filming process, and you should be willing to devote some energy into putting them at ease. (One suggestion here is to simply audiotape the most extremely camera-shy and loop their voices over other images. Voices over other images, in general, is to be encouraged at all costs.)

Sometimes the **setting** is as much important as the characters. Don't forget that this is a visual medium, and no matter how much your subject may have to say, the physical details of his environment will have a lot to do with the finished story. This might be a good place to think about **B roll**—footage of the environment which is not dominated by verbal content.

Even though it's great to get an interview right in the middle of the process you're dealing with (because it might bring out the best in the witness) the sheer level of noise associated with a lot of the most interesting environments suggests that just getting plenty of visual coverage of that environment, documenting the processes involved as vividly as possible, might be preferable. Then you can get the people in a better sound situation (don't forget that we have soundproof rooms and even a green screen room in the Media Depot of the Young Library) and overlay good sound and good images. But, with maybe a few exceptions, the images are vital to the form you are working with.

A **crew** in this case might be just a partner to watch the camera and sound while you are interviewing, but it helps a lot, no matter how small or short the production, to have another set of eyes on the recording of the event while you attend to the event itself. You don't want to mess up full concentration on either the interview or the equipment by doubling up on both. On the other hand, too many people around the edges can be a real distraction to cast and crew both, so keep things to a minimum.

You are asking people for valuable slices of their personal time, so it's important to pay attention to **scheduling.** Probably your project will not require an inordinate amount of synchronization, but always allow more time than you think you will need for interviews, which might prove much more fruitful than you expected, or which might be interrupted by unforeseen factors like weather or equipment problems.

Also, be careful about backing up interviews too close to each other, or not allowing enough breakdown and travel time between them. Even the most cheerful and cooperative witness is going to become disgruntled and even hostile if put on hold for too long.

Now you will need to do some **research.** How much? This varies according to your subject matter and how much you already know about it (which, believe me, is never half as much as you think you do). But you can basically never be over-prepared. And herein lies a paradox: on the one hand, you'll want to have as much information available as possible, and you'll need to be armed with more thoughtful, nuanced, open-ended questions than you'll probably ever be able to use.

On the other hand, you want to be able to abandon all of this on a second's notice when a new vein of information/opinion/anecdotal evidence opens up which you had never considered—and many times, it will. But you can't expect to fly by the seat of your pants—if you don't feel confident with the preparation you have already done, you won't be able to knowledgeably and competently improvise when new revelations turn up.

Another way to look at this: You want to be over-prepared, but you don't want the interview to feel like it. If you are fumbling with your notes or reading off prefabricated questions in a rote manner, you will be responded to in kind—off the cuff, glib answers, or robotic boilerplate rhetoric.

Either of these is a death sentence to your documentary. In terms of the interview, there are two extremes to avoid. One is trying so hard to look like an expert that you either alienate or intimidate the witness, either way obviously undercutting not only the interview but probably the whole vibe of the witness when your intended audience encounters them.

The other is looking—as opposed to coming across as well prepared, open-minded, and professional—so bent upon a particular thesis that the witness (and later, the audience) decides you are so convinced of your own conclusions before the investigation has even begun that there is no room left in such an airless argument for any other way to look at it. Even the scent of de facto reasoning like this will contaminate the entire argument.

The best scenario of all is to have a sequence of prepared questions which you never have to refer to. As we will discuss in the next section, the goal—for the interviewer, for the witness, and ultimately for the audience—is to feel like you're in the middle of the liveliest and smartest *conversation* anybody has ever had about this topic.

TWO: THE SHOOT

Don't be late. On the other hand, particularly if you're setting up in someone's home or business, don't get there too early either. You want to allow time to do some pre-interview warm-up while your crew gets the scene set, but it helps to have at least a pretty good idea of how tight time is and how you can make the best of it.

If you have a chance to do a pre-interview, talk about anything *except the subject of the interview.* The idea here is to develop a relationship with the witness, establish a comfort zone, and get them situated for the camera. You do not want them to wear out their mojo before you're ready for it. However, at least the impression of a pleasant relationship is worth the world, especially if the witness is not experienced in this situation. (If they are experienced, you may have the opposite problem, which is not so much teasing it out of them, but trying to steer them at all.)

On the technical side, choose a quiet and reasonably uncluttered area to stage the interview. Sound is more crucial than anything in this case: If we can't hear what the witness is saying, no amount of ingenious visuals will save it. Make that your first priority, and be sure to check and double check it before you start, both position and audibility. Check one last time with the actual witness and their individual tone. (At the end of the interview, ask for a moment of silence which you can record for ambient sound if you need it.)

Frame the shot in terms of the **rule of threes.** If you think of the space you have to work with inside the frame as a tic-tac-toe board broken into nine squares, consider placing your subject—not to the absolute extremes—but not exactly in the middle either, just a bit aslant one way or the other. Too much symmetry can be fearful. Then you might fill in some of the negative space opposite the witness with a flower, books, something to balance it out. But if it looks absolutely centered and symmetrical this subconsciously bores the viewer.

Again, put the camera where it enhances this composition, but not where the witness is looking right at it. You want him to forget about the camera if at all possible. Also, if you maintain constant eye contact throughout the interview, this will keep him where you want him, while—crucially—maintaining the intensity of everyone's engagement in the conversation. (This means, first and foremost, being well prepared enough to not have to break away to fumble with notes.)

Regarding the placement of the camera: I highly recommend placing whatever you are using—camera, iPhone, tablet, whatever—in a stable, stationary position for the duration of the interview. Many first time documentary makers have romantic illusions of "going hand held," and this approach simply causes many more problems than it solves in interviews. The time for camera movement is during the B roll shooting, when you're not concerned with sound. Even there a little bit of shaky goes a long way. You want the interview to be as stable and smooth as possible, and that means keeping the camera still.

If your subject does not seem exceptionally nervous, discuss this next issue before you start. Even though you want the interview to be a conversation, people tend to drop their antecedents a lot when they speak informally, and this can play hell with you in the editing room. So, for instance if the question is "What color were Warren Oates' eyes?" you might get a beautiful poetic response like "Blue. The color of the blue-grass, the color of the sky. The bluest blue you ever saw." Great stuff, except nobody will know that *Warren Oates'* particular eyes were the subject of that sentence. So if they can just back up and say "Warren Oates' eyes were blue. The color of bluegrass etc." it will be much easier to weave in smoothly to the final edit.

(If they seem overly anxious anyway, though, this advice will just make them more so.) If you don't have that establishing context, the alternatives are to leave in the question (which I am totally against; the only time you should show yourself or leave in your question is in special instances when it has an unusually heightened emotional impact) or use a graphic, which can work sometimes but often just looks amateurish.

Digressions can be tough to call. If the witness gets way off your original subject, think quick and ask yourself: Are they so nervous that I need to let them get a little of this out of their system? Are they so off track that I need to step on them immediately to get back to topic? Or, sometimes: Yes, this isn't what I asked but it's great stuff and I need to think about how to follow up this new development instead of trying to drag it back to my next preconceived question. Digressions can ruin an interview, or make it truly great. And you will need to make the call on the spot.

Be sure to thank the witness at the end. I always make a point of asking if there was something they really wanted to say which didn't come up in the questions. After that moment of ambient sound, be sure to ask them if they are aware of any resources—be it other people, photos, video, etc.—which might be useful to your project. You would be amazed what this will generate sometimes.

Now get all the B roll you can. What is the environment like? What kinds of processes are going on? Are they paintings or posters on the walls you might want to use for a close-up or a cutaway? If there is a natural environment involved at all—woods, plants, trees, animals—be sure to capture both the sights *and* the sounds of it as much as possible.

You're ready for the most challenging part of the process of all.

THREE: POST PRODUCTION

Now the wild, wild work begins. Prior to this, you have established the parameters and focus of your subject and shot (most of, probably) the raw footage. Now it's time to arrange those pieces of the puzzle into a compelling sequence. Before the actual edit begins, however, there are at least of couple of more details to consider.

You do *not* want the film to be an unvarying succession of mid-range interview shots, an uninterrupted parade of what are rightly derided as "talking heads." We have all grown accustomed to such a wide range of visual stimuli that the talking heads will be a lot more compelling if we've got a variety of images to look at while they are *doing* at least some of that talking—photos, selfies, headlines, graphics, B roll footage and such moving images footage as home movies, film clips, and even stock footage.

So collect as much of this as you can. It will be a crucial layer to the finished product. Ideally, a lot of it will come from research and your interview subjects. But there are plenty of links to archival sites which you are freely encouraged to use.

This goes double for music and leads us to the question of copyright. Follow your instructor's lead on this one. Personally, after dealing with intellectual property problems for more than twenty years, swimming against the vast and murky tide of unions and agents and lawyers who protect (and exploit) the increasingly arcane details of ownership and ancillaries, it seems like such a soul-killing process that I encourage you , as students and under the legal exemption for student work (which must be unpaid for and unpublished), to grab it when and where and while you can get it. While you are protected, my argument goes, why not benefit from working with the very best and most particular materials available?

But ... you can't profit from it, (don't worry too much there; the profit margin on documentary work is pretty much a sad punchline anyway). And your instructor might insist you generate your own material, partially on the argument that facing these inevitable realpolitik practicalities makes for a better learning experience.

Either way, you'll need to think about how to handle narration, i.e., the process of giving the viewer an overview and a context for what she is seeing/hearing. Cinéma vérité has proven that it can be done with nothing more than sensitive and nuanced editing—and even though I write a lot of voice-over, I absolutely agree that on some platonic level that is often the ideal—but it's a mighty hard thing to execute.

Excise, at all costs, anything which even momentarily snags up your viewer. She needs to know where she stands, because in five minutes it's going to be over. And—especially when the rough draft edit is done—you're going to spot holes where context or clarification or just a single telling detail are needed.

Subtle use of **voice-over** is one solution, but consider what voice-over *shouldn't* try to do: point out the obvious, berate the viewer before she can begin to develop her own reactions, transparently try to manipulate her reaction. Too much pathos and a killing blow to ethos, a laundry list of facts and statistics which no one will remember minutes later.

A note on logos: Logic, and a logic you develop for yourself, will be crucial to your documentary argument, but sheer numbers, in particular, what Mark Twain calls "lies, damned lies, and statistics" are almost impossible to make stick in a short film. Better—almost always—to let the particular and the local speak for the abstract and the universal.

Instead, good VO should provide a context, subtly cue us for what we should be anticipating to see and hear, and—most importantly—set the tone for the documentary, supplying a lens through which we can at least believe we are drawing our own conclusions.

Another option is: text or graphics. A little bit of either goes a long way. Just as you want to exclude your own questions from the video, repeating every question in text gets old fast. Better to employ a few "chapter headings"—a la Tarantino—maybe just a work or two which establishes the context without over-explaining it. Less is more. I am almost always against large blocks of text on film because I hate trying to read too many words on a screen. That's what books are for.

As for graphics, there are all kinds of groovy little gizmos on most editing software that reverse the shot, or polarize it, or frame in in a cute geometric pattern, etc. *Eschew* these at all costs. They are the mark of an amateur who has just found a new toy. A straight cut or—if you need a pause for dramatic emphasis—a standard dissolve is usually plenty. Having a "hand drawn" or cartoonish look, or even brief bursts of animation—especially in credits or visualization of a complex idea—can add a lot, but only if it looks like You Did It Yourself and weren't just randomly clicking apps.

Now let's think about the rough cut. At every level, WRD is all about the idea that any and all forms of discourse constitute writing, whether it be a chapter or a blog or a text or … a film. Visual rhetoric is rhetoric just the same. But again, there is are useful distinctions, comparisons and contrasts underneath this larger notion of *composition.*

Think of it this way: You are *building up* a chapter, generating words and moving outwards. Whereas, if you've done your homework on the pre-production and the actual shoot, you should have such a wealth of resources, an embarrassment of riches, that it's much more a matter of a process of *elimination* of honing down, of finding the central ideas and sculpting out the most effective components.

Try to think of a rudimentary beginning, middle, and end, while leaving open, only temporarily, the idea that any of these could be swapped out if necessary. (And if the middle is a bit of a muddle, fear not. It's almost always that way, and if you've got some bookends on either side you can always return to it.) Here's the thing: Assemble the interview "beats" by content (it might even help to work off a written transcript or just close your eyes and listen to it like the radio).

Then cut each of these elements longer than you expect them to end up, knowing that the second run through will be plenty of time to start trimming. For now, leave plenty of room on either side of the edit, and make an initial assemblage of these first chunks which will serve not so much as building blocks, but, in some instances, as scaffolding which can be removed after the building is completed. Odds are good it will run two or three times longer than your final cut.

Now start whittling it down from there. Subtract what is not essential.

Think of this first edit as the bottom layer of a cake you are building upward, into more and more narrow slices. Weave your characters (because that's what they are, and it behooves you not to use so many that they only appear once or there is no sense of personality developed) in and out so that they can tell their part of your story.

Eventually each witness will need, at the very least, what is called a **lower third,** text (usually but not always at the bottom of the frame) which spells out their name, title (if needed) and their particular ethos in terms of this rhetorical situation: i.e., "Tom Marksbury, film historian." You may just need the one tagline/ID for a short film, but if it's been a while since we've heard from this person, repeat.

The next layer of the cake—after you've *tightened*, every time through, the longer cut to your increasing satisfaction—is dropping in the archival material. Look for two primary opportunities here: 1) any time you want to break away from the talking head—usually best to let somebody either start their comment and cut away so that he finishes it as a disembodies voice over another image, or 2) vice versa, where we are looking at the illustration of the comment and then return to the witness for the dramatic finish.

Some documentaries are stark, relentless, almost fixated on certain images. Others are dense (sometimes too dense) with constantly fluctuating patterns of images which almost dare you to look away. Both your content and your own temperament will dictate which approach you take—and there is certainly room, within reason, for a mixture of both. Always remember that these suggestions are all only *alternatives,* and that you should never try to please everybody by throwing every possible alternative in. *Construction* means building it up; *editing,* in the end, means scaling it back and cooking it down.

The other issue to consider—let's call it "covering over"—involves jump cuts. Nobody speaks in perfect paragraphs—and you probably don't want verbiage that textbook like in your film anyway. So you want to preserve their conversational style, but all of us have little tics and tricks—"umm," "like," etc.—we habitually use to buy time to put our next piece of thought together.

Some of that is fine, but a lot of it can be a real distraction. So you may want to make a tiny cut for better continuity and flow. *Oral* continuity, however, is inevitably going to interrupt visual continuity, so sometimes where that little slice is missing you might want to paper over the cut with a visual—photos, cartoons, etc. Sometimes jump cuts are not "accidents" at all, but are used deliberately to provide an edgy, exciting style (see Godard or way too many music videos). A little of this goes a long way; don't overuse them.

Along these same lines, consider the impact of the long unbroken take versus lots of very quick, even frenzied, sometimes nearly subliminal editing. Both have their place, but too much of either can be really tedious and/or headache-inducing. If a subject builds to an especially extended emotional or impactful moment, you don't want to interrupt it. Keep in mind, too—not only when you're editing but in the initial interview itself—to allow a moment *after* a particular high point (in other words, don't step on it at the time or cut it too short later). Sometimes the subject will sigh, or smile, and add some little bit of body language which greatly expand the impact of what he "just said."

On the other hand, probably for better *and* worse, we've all come to expect hyper-editing which forces a number of images (and remember the first rule of montage sometimes dictates not just a series of similar images but a collision of violently contradictory ones) to wash over us. Again, this is excellent in short spurts, but only to a point, and certainly not unabated. Too much of it, just like too much of any of these other tactics, and the mind simply shuts down.

The last layer is the voice-over, if you decide to use it, and the music. Again, if you want to use *any* music you find particularly appropriate, fine with me. But your instructor may want you to consult one of the many websites where you can find uncopyrighted generic music for free. Or better yet, for you musicians out there, why not compose and perform your own soundtrack music? It worked for John Carpenter in *Halloween.*

Again and again, though, *less is more.* You don't want overly emphatic, even melodramatic music stepping on every lovely point you've already made perfectly well, and you don't want a nonstop wallpaper of back to back music through every second of the film. It is *not* a music video. No need to include that entire guitar solo because it's always been your favorite—and the last thing you want to do is drown out the speaker. Unless a short burst of lyrics is unusually relevant to the proceedings, instrumental music in general is less distracting and will cause you fewer sound mixing problems.

Speaking of—once you're close to a final cut, try to adjust the soundtrack audio so that all sounds are reasonably consistent. Nobody is expecting Academy Award quality technical work on any of these issues, but it's very important that the sound—the audible end of the story you are telling visually—be reliably audible and more or less consistent.

That about does it. Don't get too fancy with the credits, but include some, and don't forget to document everything you've borrowed, the same as you would on the "Works Cited" page of a chapter.

Give it a good title, and put the title up early. The credits should come at the end of such a short film, otherwise it seems like a very pretentious instance of the cart before the horse. But in terms of cuing an audience who are only going to be spending a very small amount of time with your film what to expect (and what not to expect, and what to continue to think about after the film is over and beginning to resonate), a bold, provocative, and short tile can go a long way.

All right, then. You've got it. Hard work but more fun than you expected, right?

Now: Repeat. Do it again.

BIBLIOGRAPHY

As with screenwriting, there are way more books and software on the market than you will ever find useful. The best way to learn is just to grab a camera and get out there. I did want to mention a few solid nuts-and-bolts approaches, and a couple slightly more philosophical overviews. Many of the technical problems which may arise can be investigated on YouTube.

NUTS AND BOLTS

The Shut Up and Shoot Documentary Guide, Anthony Artis, Focal Press, 2008.

- Every bit as basic as it sounds. Covers a lot of territory in very direct and succinct terms. Absolutely jargon free, but with a useful glossary and lots of visual illustration. Almost a total focus on technical problems, not much on process.

Writing, Directing, and Producing Documentary Films and Videos, 4th edition., Alan Rosenthal, Southern Illinois University Press, 2007.

- Much more info on process here—Rosenthal is very good at leading you through the questions you need to be asking both way before you shoot and way after you have all the raw footage you need. There's a little too much focus on feature length films, thus you will find sections on working with film stock, budgets, and grant proposals, more than you will probably need. But for navigating through the steps and underlining what you need to be concentrating on, even for a very short film, a very useful text.

Documentary Storytelling: Creative Non-Fiction on Screen. Sheila Curran Bernard, Focal Press, 2011

- Very helpful emphasis on narrative as the backbone for any kind of film, non-fiction or not. She can help you structure your material, think about pacing and exposition, what to emphasize and what to exclude, etc. Again, this veers a little more to the conceptual, but her storytelling instincts are very sound, and she can help you get in better touch with yours.

SLIGHTLY MORE AESTHETIC/PHILOSOPHICAL (FOR STUDENTS WHO WANT TO DIG A LITTLE DEEPER)

Introduction to Documentary, Bill Nichols, 2nd edition, University of Indiana Press, 2006.

- Probably the most incisive and comprehensive thinking about the form in terms of both theory and application. Nichols devotes a chapter each to nine essential questions, from "what makes documentaries engaging and persuasive?" to "how have documentaries addressed social and political issues?" which he proceeds to answer to both very practical and extremely philosophical terms.

Documentary: A Routledge Film Guidebook, Dave Saunders, (Routledge Press,2010)

- Tends to concentrate on more recent work, which might be more useful to first year students who might be otherwise overwhelmed by the vast history of the documentary form. Special attention to experimental work and marginalized voices.

Documenting the Documentary: Close Readings of Documentary Films and Videos, edited by Barry Keith Grant and Jeanette Sloniowski, Wayne State University Press, 1998

- Excellent critical analysis of canonical films from *Nanook of the North, Man With the Movie Camera,* and *Triumph of the Will* to more recent classics like *Grizzly Man* and *Four Little Girls*. Particularly useful for more advanced students who might want to write *about* documentaries.

EVEN MORE PHILOSOPHICAL (BUT ESSENTIAL IF YOU REALLY WANT TO KEEP DOING THIS)

Ways of Seeing, John Berger, Penguin

On Photography, Susan Sontag, Picador

Both these books are both extremely philosophical—almost metaphysical—but very practical at the same time, the best of all possible combinations. Berger is of great use to anyone trying to figure out any kind of visual image in any kind of frame, and his discussions of such issues as perspective and point of view will benefit not only the casual art fan but, crucially in our case, anyone trying to compose and frame up the images he is trying to compile for a documentary.

In a similar vein, Sontag is really talking about the moral implications of the choices we make when we photograph (violate? validate? Impose our own unconscious set of values upon?) Anything up to and including ourselves. You don't want to be a cultural tourist, and you want to take responsibility for your own agenda—which first and foremost means realizing that you have one) and Sontag can straighten you out on that right away.

A FISTFUL OF DOCUMENTARIES

My two documentary production classes (WRD 312: Introduction to Documentary and WRD 412 Intermediate Documentary) both feature a wide variety of works which stress an even wider range of structural, expositional, and narrative techniques. I also teach a History of Documentary course (WRD 311) which traces the evolution of the form from the Lumiére Brothers at the turn of the twentieth century to ... well, yesterday. What follows, though, is a very short and selective list of films which might inspire students at the first year Composition and Communications level.

Somewhat Experimental

Grizzly Man (Werner Herzog, 2005) 103 m

Not only a compelling portrait of Tim Treadwell, who dedicated his life to grizzly bears and was eventually mauled to death by one, but a fascinating subtext as Herzog, well known for his own completely antithetical obsessions with nature, wrestles with his own subjective point of view while trying to assemble Treadwell's footage in ways the activist could never have predicted or intended.

Exit Through the Gift Shop (Banksy, 2010) 87 m

Again, a dance of focus between the ostensible "director" (Terry, who may or may not be real) and "subject" (Banksy, ditto). To call this a "mockumentary" would be to sell it short by half. It's almost like crawling through a hall of distorting mirrors, while learning a lot about commercial and outsider art along the way.

This Is Spinal Tap (Rob Reiner, 1984) 82 m

The mockumentary actually goes all the way back to *Haxan: Witchcraft through the Ages*, from silent film days, but this evisceration of "Behind the Music" clichés, following the pathetic and vainglorious reunion of a third rate hair metal band, not only illuminates the connection of insecure teenage boys to preening cockrockers, but explodes most of our most naïve assumptions about what documentaries can and can't do along the way.

Cinéma Vérité

Salesman (Albert and David Maysles and Charlotte Zwerin, 1968) 84 m

As close to pure vérité as you can get. Seemingly without any directorial mediation, we travel with four itinerant Bible salesmen as they ply their wares across the country. What seems to start out as a fairly conventional morality play (the men are preying upon and profiting from the guilt and naïvité of their lower working class Catholic customers) becomes a multi-faceted study in what Thoreau called "lives of quiet desperation" and the cracks in the American dream.

Hoop Dreams (Steve James, 1994) 171 m

An epic masterpiece of socioeconomic reportage, as we follow two aspiring basketball players from inner-city Chicago as they move from their hopeless neighborhoods to scholarships in the white suburbs, trying for the big prize of a professional career. Again, the ostensible subject keeps splitting into more and more layers as we watch the sad and glorious trajectory of their efforts to get out and make it.

Engaged History and Activism

When the Levees Broke: A Requiem in Four Acts (Spike Lee, 2006) 226 m

The Hurricane Katrina story, told through a rich vein of resources including news footage, interviews with regular citizens and pundits alike, vivid graphics, and a haunting sense of growing quiet outrage you might not expect from Spike Lee. More even-handed than you'd think, and infinitely more effective because of it. (In marked contrast to Michael Moore.)

The House I Live In (Eugene Jareck, 2012) 108 m

A history and critique of the failure of the so-called Drug War, at once personal, even intimate, and, for once, a sweeping and persuasive use of statistics and numbers. Jareck manages to find the common line where disenchanted police who don't have time and supplies to do their real jobs and activists like David Simon (creator of *The Wire*) agree about this failure in terms of wasted money, wasted lives, and sheer human futility.

The Thin Blue Line (Errol Morris, 1988) 101 m

Created by America's greatest documentary maker (in my opinion) and both a life-changing experience and an argument that both the true crime genre and the artful use of re-creations don't have to be the kind of law and justice porno we watch on reality television shows like *Cops*, etc. Randall Adams, the subject of the film, was wrongly convicted of murdering a police officer. Based on the impact of the documentary, Adams was released; he subsequently sued Morris for the rights to his story.

AUTOBIOGRAPHY

Stories We Tell (Sarah Polley, 2013) 108 m

A young actress becomes obsessed with re- (and de-)constructing what she thought she understood about her childhood after the death of her mother. Almost painfully intimate, and a disarming mix of family photos, home movies, interviews that might as well be called interrogations, and re-creations that you may not even be able to spot.

Tarnation (Jonathan Caouette, 2004) 88 m

Supposed made for $200 on a laptop computer, this is one of the most harrowing and confessional films I've ever seen—the director's difficult and abusive childhood is examined in excruciating detail—and yet there is not an ounce of self-pity to be seen. Far from it, after basically spending ninety minutes inside this guy's head and history, you feel an affirmation of the human spirit. Very instructive in its innovative use of found footage, archival, and homemade graphics.

KENTUCKY

Harlan County USA (Barbara Kopple, 1980) 103 m

Near-Vérité but skillfully manipulated depictions of the coal strikes of the late nineteen seventies. The sheer will and endurance of the miners—and especially the women who support and often guide their efforts—show how little rhetoric means compared to flawed human scrappiness. A blow against the empire in general, and Duke Power in particular.

Stranger with a Camera (Elizabeth Barrett, 200).

The flip side of *Harlan County USA.* Like Kopple, the subject of this film came to Appalachia with the best of intentions, but unlike Kopple, she was unable to convince the local citizenry that she wasn't an intrusive outsider. A meditation on both the tremendous moral responsibilities and the inherent dangers of documentary film making.

Buy the Ticket, Take the Ride: Hunter S. Thompson on Film (Tom Thurman, 2007) 77 m

All right, I wrote it—but it's still pretty good. I like to point out two opposing directions the subgenre of films we might as well call "portrait of the artist" can go wrong—either it's a hagiography, where all the warts are surgically removed and nothing exists except the positive and adulatory, or the patho-biography, which goes beyond warts-and-all to reveal mostly all warts. Hunter had a lot of warts, yet he changed American writing. This is our attempt to explain these contradictions.

INDEX